EARTH:
THE WATER PLANET

Revised Edition

Jack E. Gartrell, Jr.

Jane Crowder

Jeffrey C. Callister

◆

A joint project of Horizon Research, Inc., and the American Geological Institute

Materials for middle-grade teachers in Earth science

This project was funded by BP America, Inc.

The National Science Teachers Association

Produced by Special Publications
National Science Teachers Association
1742 Connecticut Avenue NW
Washington, DC 20009

Stock Number PB–76
ISBN # 0–87355–083–8

EARTH: THE WATER PLANET

Table of Contents

◆Acknowledgements

Many people contributed to the planning and preparation of *Earth: The Water Planet*. Most important among these are the principal authors: Jack Gartrell of Horizon Research, Inc., in Research Triangle, NC; Jane Crowder of Pine Lake Middle School in Issaquah, WA; and Jeffrey C. Callister of Newburgh Free Academy in Newburgh, NY.

The manuscript was reviewed for accuracy and functionality by Marilyn Suiter of the American Geological Institute; Sue Cox of the United States Geological Survey; Ann Howe, Ellen Vasu, and Norman Anderson of North Carolina State University; Frank Ireton of the National Earth Science Teachers Association; and Walter Martin of the University of North Carolina.

Sondra Hardis of British Petroleum America, Inc., provided many suggestions that helped make this a valuable resource for teachers and helped establish links between this material and many potential users. Phyllis Marcuccio of the National Science Teachers Association handled the arrangements that made production of this book possible.

This revised edition was made possible through the efforts of Jane Crowder and Jeffrey C. Callister, who used their extensive classroom experience with the original book to update and revise the activities and to expand the suggestions for further study.

Earth: The Water Planet was produced by NSTA Special Publications, Shirley Watt Ireton, managing editor; Ward Merritt, assistant editor; Elizabeth McNeil, editorial assistant. Elizabeth McNeil was NSTA editor for *Earth: The Water Planet*. Illustrations were created by Sophie Berkheimer and Max-Karl Winkler. The book was designed by Sharon H. Wolfgang of AURAS Design.

Earth: The Water Planet was a joint project of Horizon Research, Inc., and the American Geological Institute. Project Director was Iris R. Weiss; Project Co-director was Andrew J. Verdon, Jr.

Special thanks go to British Petroleum America, Inc., for providing the funds to make this project possible.

OVERVIEW

Earth: The Water Planet

When viewed from space, Earth appears as more of a liquid planet than a solid planet.

When we use the term "water," we are usually referring to water in its liquid state. Water also exists as a gas (water vapor in the atmosphere) and as a solid (ice in glaciers and polar ice masses). Water is one of the few substances that exist in three different states at temperatures routinely encountered on Earth.

Earth: The Water Planet examines water: its scarcity or abundance, where it is found, its unique physical properties, how it moves through the atmosphere, and how it reshapes the solid Earth. This book will show you and your classes ways of examining problems resulting from the limited amount of water available, including how water affects people's jobs, the necessity for water treatment and water management, and the difficult political questions that sometimes arise over water allocation.

◆Organization of this book

Earth: The Water Planet is divided into five modules on different aspects of water. Each module includes activities that use a variety of instructional methods: hands-on experiments that challenge you to purify "swampwater," determine the rate at which soil absorbs water, and test a method of reducing erosion; conservation-oriented activities that show how much water can be wasted by a dripping faucet, and how much we depend on modern plumbing; and group activities where special interest groups of water users testify at a town meeting. Audiovisual materials are integral parts of several of the activities. Several segments of the Eureka! video series, produced by TVOntario, and selections from the AGI/ Encyclopaedia Britannica Basic Earth Science Program films are used for concept presentations.

Following the modules is a collection of readings. This section, drawn from many sources, gives detailed explanations and provides additional examples of many of the concepts explored in the workshop activities. These readings can be used to give the teacher some background on concepts to be explicated in the activities, and they can also be reproduced as student handouts to augment student discussion at the completion of lab activities.

A guide for teachers and workshop leaders is provided for use in planning instruction of *Earth: The Water Planet* activities. This section lists the equipment and materials required to perform each module's activities. Ordering information for the recommended audiovisual materials is also included in the guide for teachers and workshop leaders. Finally, a glossary provides definitions of many of the terms used in these modules. This glossary also serves as a master list of vocabulary introduced and defined in the activities.

◆Getting ready for classroom instruction

Most of the activities in this book can be adapted for classroom demonstrations. The authors have tried to make the material as close to life as possible to enhance student acceptance and enjoyment of the activities as activities, however.

Each module begins with a discussion of the rationale, objectives, and overview of the activities within it. Each activity begins with a reproducible student worksheet, which also contains the concept objective and any new vocabulary words. Each activity is designed to take between one and two class periods (40–60 minutes) to complete.

Each activity worksheet is followed by a commentary for teachers called "Guide to Activity...." These sections explain the expected results for the activity and provide additional background information on the content of the lesson. You also will find suggestions for time management, teacher preparation, ways to help students obtain reproducible results, ways of troubleshooting equipment, and hints about problems that may be encountered while performing the activity. Notes on safety, sample data, and answers to the questions on the activity worksheet are also provided. In addition, these commentaries contain suggestions for further study, outlining other experiments that can be performed using the same apparatus.

Good results can be obtained for many activities by using alternative procedures or by using substitutes for materials listed on the worksheet. If you do not have ready access to the materials listed, the "Guide to Activity...." section offers possible substitutions and procedure modifications.

◆Getting ready for workshops

If you are using this book as a workshop leader or teacher participant, you may wish to perform some of the activities as classroom demonstrations. As you work through these activities with your colleagues, you will have the opportunity to discuss new insights, explore alternative procedures, and make note of any problems you encounter. These experiences will add to your confidence when you direct your students in similar activities later.

Each module is designed to take one inservice workshop time period. The general topic of each module is described in its introduction. Each introduction lists instructional objectives for the workshop, gives the titles of the activities in the module, and indicates the readings that should be studied after the module's activities are performed.

MODULE 1

Groundwater—The Largest Freshwater Resource in the USA

◆Introduction

• Earth's supply of fresh surface water held in lakes, rivers, and streams is only about 3% of the amount lying underground. If all the world's surface water were combined into a single lake, that lake would contain enough water to cover Indiana to a depth of about 1500 m (almost a mile deep). That would be a big, deep lake, but it would be small in comparison to the lake that could be formed from all of Earth's underground water.

• Hydrologists, scientists who study Earth's water resources, estimate that there is enough fresh groundwater lying within 800 m (half a mile) of Earth's surface to form a lake that could cover the states of Texas, New Mexico, Colorado, Oklahoma, Kansas, Nebraska, Wyoming, South Dakota, North Dakota, and Montana to a depth of about 1500 m.

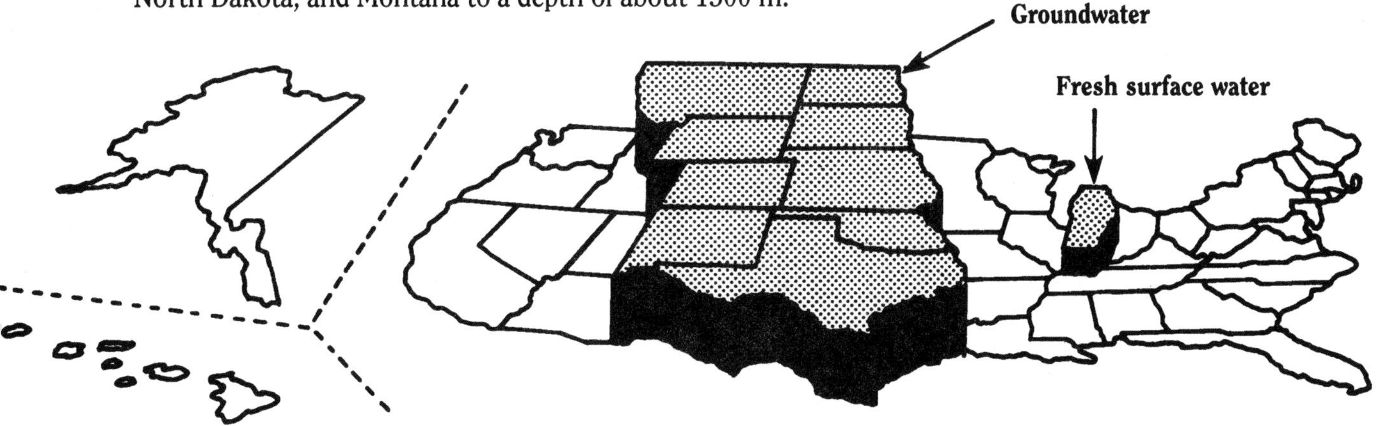

This module deals with groundwater and issues relating to the use of groundwater. During the workshop you will perform several hands-on activities that demonstrate how water is absorbed by the soil and moves underground. You will also participate in a role-playing exercise that examines how the careers, homes, and life-styles of the members of a community might be affected by problems related to groundwater shortages and groundwater pollution.

Before we begin to investigate groundwater, we need to define some terminology that will be used throughout these modules. Do not be

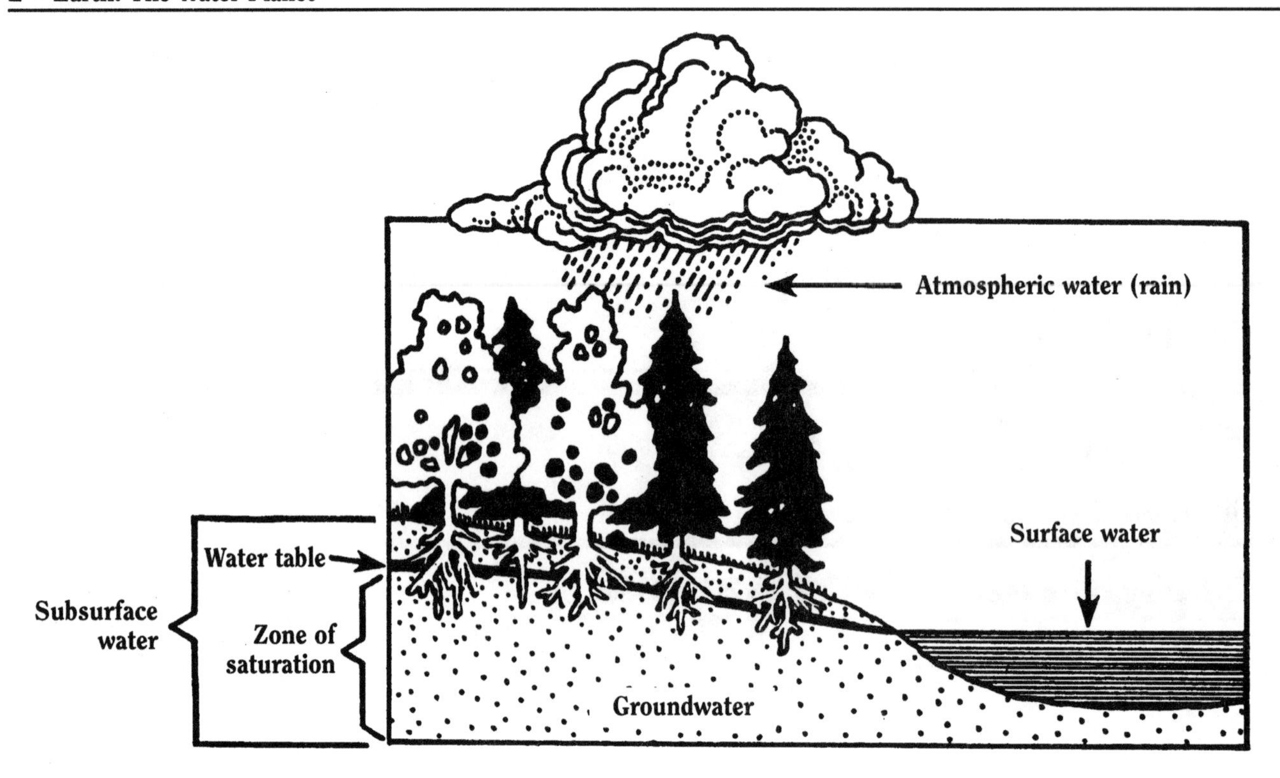

concerned if these terms are unfamiliar to you. You will become more comfortable with them as you do the hands-on activities and read the associated materials. We have also included a glossary of these terms at the end of these modules.

Scientists use the term **surface water** to refer to all water, solid as well as liquid, on the surface of the Earth but below the atmosphere. **Subsurface water** is all water, solid as well as liquid, below the Earth's surface.

All soil, even the soil in deserts, contains some water. In temperate regions, plants absorb a large amount of the water entering the soil; gravity moves the remaining water downward. The water eventually reaches the **zone of saturation**—a level underground where the soil and rock contain as much water as they can hold. All the spaces between the rocks and soil particles are filled with water; we call this water **groundwater**. The top of the zone of saturation is called the **water table.** Digging or drilling a hole into the zone of saturation creates a space into which water will flow. We call such holes wells.

The amount of water that underground rock or soil can hold depends on its **porosity**—the ratio of open space to the total volume of the rock. In sand or gravel **aquifers** (water-bearing rock layers), the porosity may be 30% to 40%. In some types of rocks, or sediment layers, the porosity may be only 1% or less.

Most groundwater is in motion. Gravity pulls it downward. Since the water table is not perfectly flat, and since water sinking into the ground pushes against the water that is already underground, groundwater also flows in a horizontal direction along hydraulic gradients that reflect the slope of the water table or the pressure of water within an aquifer.

Groundwater does not reach all the way to the center of the Earth. Layers of watertight rocks lie at depths ranging from 100 to 10,000 m. No water can flow beneath these watertight layers.

You will note that these modules use metric units wherever possible. These are the units routinely used in science. Length is typically measured in meters (m). One meter is equivalent to 39.37 inches, or slightly more than a yard. Shorter lengths are measured in centimeters (cm) or millimeters (mm). One meter is equal to 100 cm or 1000 mm.

Vocabulary

- **Aquifer:** An underground layer of rock, sediment, or soil that is filled or saturated with water.
- **Porosity:** A measure of the ratio of open space within a rock or soil to its total volume.
- **Subsurface water:** All water, solid, liquid, or gaseous, that occurs beneath the Earth's surface. Subsurface water is groundwater located below the water table in the zone of saturation.
- **Surface water:** All water, fresh and salty, on the Earth's surface. Oceans, lakes, streams, snow, and glaciers are all surface water.
- **Water table:** The boundary in the ground between where the ground is saturated with water (zone of saturation) and where the ground is filled with water and air (zone of aeration).
- **Zone of saturation:** The portion of the ground below the water table where all the pores in rock, sediment, and soil are filled with water.

Large distances are measured in kilometers (km). One kilometer is equal to 1000m, or about 0.6 mile.

Volume is typically measured in liters (L). One liter is slightly more than a quart. A gallon is equal to 3.8 L. Small volumes are usually measured in milliliters (ml). 1 L is equal to 1000 ml. One teaspoon of a liquid is equal to about 5 ml.

In many cases the English system equivalent is provided to help you and your students make the translations. We have also included a metric conversion chart at the end of these modules.

SUMMARY OF MODULE 1

Groundwater—The Largest Freshwater Resource in the USA

◆Instructional objectives

After completing the activities and readings for Module 1, you should be able to

- demonstrate how water moves through the rock and sediment of an aquifer [Activity 1]
- measure the rate at which soil absorbs water [Activity 2]
- demonstrate the presence of pore spaces in sediments and rocks [Activities 2 & 3]
- show the relationship between soil composition and its ability to hold water [Activities 3 & 4]
- show why having a high-quality, reliable water supply is essential to the economic well-being of a community [Activity 5]

◆Preparation

Study the following readings for Module 1:

Reading 1: Groundwater: What, Where, How, and Why
Reading 2: The Ins and Outs of Groundwater
Reading 3: Quality and Protection of Our Groundwater
Reading 4: Managing Hands-on Activities in the Classroom

◆Activities

This module includes the following activities:

Activity 1: Can Water Move Through "Solid Rock"?
Activity 2: How Fast Does Soil Absorb Water?
Activity 3: Is It Full Now?
Activity 4: How Much Water Can Different Soils Hold?
Activity 5: Who Is Responsible for Preserving Anyville's Water Supply?

ACTIVITY 1 WORKSHEET

Can Water Move Through "Solid Rock"?

Materials

Each group will need

- a dry, unglazed clay flowerpot
- a disposable metal pie pan
- water
- a metric ruler
- a large paper grocery bag

Vocabulary

- **Capillary action:** The process by which water rises through rock, sediment, or soil. The cohesion between water molecules and adhesion between water and other materials "pull" the water upward; also called capillary action.

◆Background

Water is constantly moving through "solid" underground rocks. In this activity, you will use a clay flowerpot as a model to demonstrate that water can move through solid material by **capillary action.**

◆Objective

To demonstrate how water moves through the rock and sediment of an aquifer

◆Procedure

1. Your teacher will provide you with a clean, unglazed clay flowerpot that earlier was placed in an oven set at 150° C (about 300° F) for 30 minutes. Heating removes the moisture that is trapped in the pores of the flowerpot.

2. Pour water into the metal pie pan to a depth of 1 cm (measure this depth using a metric ruler). Set the dry flowerpot in the water. Avoid wetting its top and sides when you place it in the water. Record the time that you placed the flowerpot in the water in the data table at left.

3. Observe the flowerpot at 10-minute intervals.

After a few minutes, the surface of the flowerpot above the water in the pan will begin to look and feel damp.

With a metric ruler, measure the distance between the water level in the pan and this "line of dampness" on the flowerpot (measure to the nearest 0.1 cm). Record your observations in the data table at left:

4. After 30 minutes, remove the flowerpot from the water and examine it thoroughly. Describe its appearance.

____

____

____

____

Height of line of dampness

Starting time ____________

After 10 minutes ________ cm

After 20 minutes ________ cm

After 30 minutes ________ cm

5. Determining whether the water is moving *through* the clay of the flowerpot or just *on its surface* is not possible by looking at the outer surfaces of the pot. Your teacher will place one of the pots inside a heavy bag and break it, so that you can examine the clay beneath the surface of the pot.

! Some of the pieces of the broken pot may have sharp edges—handle all fragments with care!

6. Find a fragment of the pot that has *damp* surfaces. Look at the broken edge of this fragment. Does the broken edge look damp also? Does the water appear to be present *inside* the fragment as well as on its surfaces? Describe the appearance of the area exposed by breaking the pot.

____

____

____

7. Now find a fragment of the pot that is dry. Look at the broken edge of this fragment. Does the broken edge look damp or dry? Does this fragment appear to have water inside it? Describe the appearance of the area along the broken edge of the fragment.

____

____

____

8. Bricks used in constructing houses are composed of clay that is very similar to that of the clay flowerpot. Why do you think that it is necessary to coat bricks that will be underground with asphalt or some other waterproofing compound when building the foundation of a house?

____

____

____

GUIDE TO ACTIVITY 1

Can Water Move Through "Solid Rock"?

Vocabulary

• **Capillary action:** The process by which water rises through rock, sediment, or soil. The cohesion between water molecules and adhesion between water and other materials "pull" the water upward; also called capillary action.

• **Permeability:** The capacity or ability of a porous rock, sediment, or soil to allow the movement of water through its pores.

! When unglazed clay pots break, they usually do not produce fragments having sharp edges. Even though the probability of eye injury while performing this step is very small, we strongly recommend that you (and any student assistants) wear safety goggles while breaking the pot.

◆What is happening?

Rocks and soils that have a large number of interconnected pores usually allow water to flow through them quickly; such soils are said to be highly **permeable**. The flowerpot is used as a model of a permeable rock for this activity.

The results of this activity come as a surprise to many people. Most of us think that solid objects such as clay pots (our model rock) or naturally-occurring rock deposits are "waterproof." After all, clay pots can hold water without leakage through the sides of the pot. Similarly, water does not appear to pass directly through solid rock. However, closely observing what happens to a dry flowerpot placed in water shows that water *can* move through solid objects by **capillary action**. The same phenomenon explains the ability of underground water to move slowly through rock.

Porosity, which is a measure of the ratio of open space within a rock or soil to its total volume, is one of the factors that affects permeability. In general, the more porous the rock, the faster water can move through it.

Porosity is not the only factor controlling the movement of water through the ground, however. Some soils have a high porosity, but the pores are so small that friction between the water and the soil particles greatly impedes the water's motion. Therefore, water moves through these soils quite slowly; as a result, they are fairly impermeable.

Rocks such as granite and basalt may be permeable even though they have a very low porosity: if there are connecting cracks in these rocks, water can move through the cracks.

◆Time management

You may wish to give students a reading assignment to work on while they are waiting to measure the heights of the line of dampness. The actual measurements take very little of the 30 minutes allotted for the activity.

◆Preparation

At least two hours before performing this activity, dry the flowerpots as follows:

Heat clean, unglazed clay flowerpots in an oven set at about 150°C (300°F) for 30 minutes. Allow the pots to cool to room temperature before performing the activity.

Similar results can be obtained for this activity by drying the flowerpots several days in advance and storing them in a dry area until using them. Wrap the pots in airtight plastic bags if the humidity is very high or if you plan to store them for more than a few days.

A safe way to break a clay pot

Place the pot inside a large paper grocery bag. Close the top of the bag by folding it over on itself several times. Put on your safety goggles.

Drop the bag containing the pot on a hard floor. (Hint: drop the bag about 1 meter. You want the pot to break into several large pieces, not shatter into tiny fragments.)

Carefully remove the broken flowerpot from the bag.

Breaking a single pot should provide enough "damp" and "dry" fragments for an entire class to examine.

Before distributing the pieces of broken pot to your class, remove any fragments having sharp edges from the bag. Alternatively, you can simply show the "damp" and "dry" fragments to your class without having the students handle the broken pieces.

! Some of the pieces may have sharp edges—handle all fragments with care!

◆Suggestions for further study

Kids love to experiment with rocks they find and bring to class. Some rocks will absorb a measurable amount of water. To test a rock for water absorption, have the student weigh it. Then soak it in water for half an hour. After soaking, dry the surface of the rock with a towel and weigh the rock again. If the rock absorbs water, the second measurement will show that its weight increased.

If pieces of sandstone are available, they can be used instead of (or in addition to) the clay pots to demonstrate how rocks absorb water; the sandstone works particularly well.

Use several different kinds of rocks such as sandstone, granite, vesicular basalt, etc., and then discuss why some rocks absorb more water than others. This introduces the concept of pore spaces used later in this module.

◆Answers

The following Sample Data are intended as a guideline only. Your data will vary from the results listed at right.

Sample data

Height of line of dampness

Starting time	*11:20*	am/pm
After 10 minutes	*2*	cm
After 20 minutes	*3*	cm
After 30 minutes	*3.5*	cm
(After 60 minutes	*5*	cm)

4. The part of the flowerpot nearest the water appears damp on both the inner and outer surfaces; the top part of the pot appears dry.

6. The broken edge of a damp fragment also appears damp. There seems to be water *inside* the fragment as well as on its surface.

7. Dry fragments appear to be dry all the way through.

8. Bricks and cinder blocks below ground must be waterproofed, otherwise water may seep into the basement or crawl space.

ACTIVITY 2 WORKSHEET

How Fast Does Soil Absorb Water?

Materials

Each group will need

- a large (1.5 L or larger) fruit- or vegetable-juice can with both ends removed (the can must be approximately 18 cm tall and 10 cm in diameter)
- safety goggles for each group member
- a hammer
- a 2 x 4 wooden board about 35 cm in length
- a metric ruler
- 1 L of water
- a watch or clock
- a grease pencil or permanent marker

Vocabulary

- **Infiltration:** The entrance or flow of water into the soil, sediment, or rocks of the Earth's surface; also called percolation.

◆Background

The composition of soil is one factor that affects the rate at which water is absorbed. The soil that we see on the Earth's surface varies in composition from the loose sand of ocean beaches to brick-hard clay exposed by highway construction. However, most soils have a composition that lies somewhere between these extreme soil types. The soil in your schoolyard probably contains some clay, some sand and pebbles, and a certain amount of plant material. For this activity you will test how quickly your local soil absorbs water. This process is called **infiltration.**

◆Objective

To measure the rate at which soil absorbs water

◆Procedure

1. Using a crayon or grease pencil, draw a mark 5 cm from one end of the can (on the outside of the can).

2. Gather the materials and equipment listed above. Take them to a typical area of soil outside the building. If possible, avoid areas that are rocky—it will be difficult to drive the can into rocky soil.

3. Before you disturb the soil in any way, describe the soil as best you can in the spaces below:

Location: (on a lawn, in woods, in a flower bed, a sandbox, etc.)

____

Plant material present: (grass, moss, dead leaves, sticks, etc.)

____

Soil condition: (Is the soil surface dry, moist, rocky, sandy, bare, hard clay, loose granular, etc?)

____

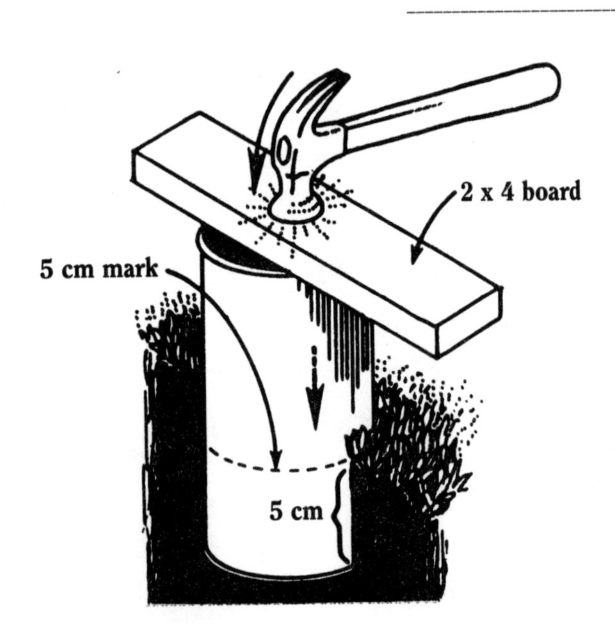

4. Put on your safety goggles. Set the can on the ground. Turn the can so that you can see the line that you drew. Place the 2 x 4 board on top of the can.
By hitting the board with the hammer, drive the can 5 cm into the ground. Use the line that you drew on the can as a depth marker.
Change the position of the board on the edge of the can after each blow. The board will spread out the force of the hammer blows so that you will not bend the can.

5. Pour 1 L of water into the can. Record the time when you poured the water into the can:

__________ am/pm.

6. Immediately make a "starting-height marker" by drawing a line with a crayon or grease pencil inside the can at the initial water level.

Using a metric ruler, measure the distance from the starting-height marker to ground level. This is the water's starting height.

Starting height : ________ cm.

! Flatten any sharp edges that remain inside the rim of the can. This will help you avoid cutting your fingers and will make the can easier to sink into the ground.

7. As the water is absorbed by the soil, the height of the water remaining in the can will decrease. You can determine how many cm of water are being absorbed by the soil by measuring the distance between the starting-height marker and the surface of the water.

Using a metric ruler, measure this distance 5, 10, 30, and 60 minutes after you pour the water into the can. Record your measurements in the data table at right.

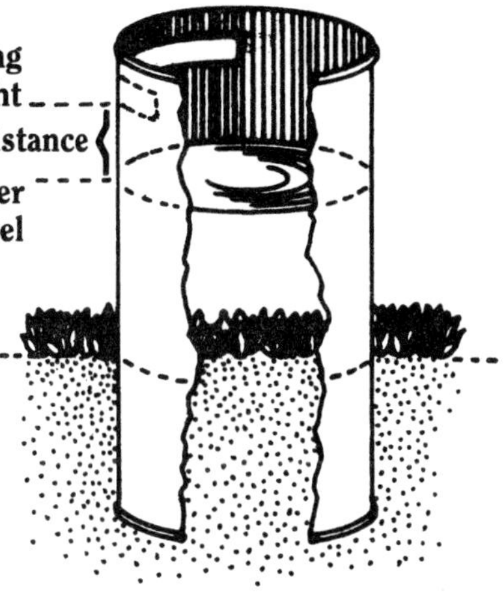

Starting height

Measure this distance

Water level

Ground level

8. On the basis of your observations, predict whether or not a large flat area of soil similar to the soil that you tested could

a) absorb 5 cm of rain striking the ground during a slow, steady rainstorm that lasts 10 hours.

b) absorb 5 cm of rain striking the ground during a heavy thunderstorm that lasts one half hour.

Explain your predictions.

____

____

Centimeters of water absorbed by soil

Time *(minutes)*	Distance between starting-height marker and the surface of the water
5	_______ cm
10	_______ cm
30	_______ cm
60	_______ cm

9. Would the predictions that you made for question 8 be different if the soil were located on a steep hillside? Explain why your predictions would change or remain the same.

____

____

10. The amount of water already in the soil also affects the rate at which water is absorbed. How do you think that the rate of absorption might change

a) if no rain had fallen in your area the week before your experiment?

b) if it had rained heavily the night before you performed this experiment?

____

____

____

GUIDE TO ACTIVITY 2

How Fast Does Soil Absorb Water?

♦What is happening?

This activity provides a baseline for the rate of water absorption by soil in your area. This value will vary, depending on the type of soil that you test. An area of lawn planted in clay soil may absorb water at a rate of about 10 cm per hour; extremely sandy, dry soil may absorb 10 cm per minute; baked subsoil exposed on an embankment may require several hours to absorb 10 cm of water.

Many factors affect the rate at which soil absorbs water. Some of the main factors are soil composition (clay, sandy, rich in organic material, etc.); soil cover (bare, cultivated, grassy, leaf-covered); amount of water already present in the soil; the slope of the land; and the rate at which rain strikes the soil (faster runoff from hillsides and faster runoff during heavy rains both reduce absorption).

In order to have groundwater, the water must have some way to get into the ground. Groundwater is created when the water in streams or lakes soaks into the surrounding soil; in some places, an entire stream may enter a cavern, becoming an underground stream. However, rainfall absorbed by the soil is the main way that groundwater supplies are produced.

♦Time management

One way to organize this activity is to have the whole class participate in setting up the equipment and making the first observation, then have the students return to class. Subsequent measurements can be made by one or two members of each group.

This activity may take longer than one period to complete. You may wish to have successive classes measure the water in the same can and combine their data the next day.

♦Preparation

Remove both ends of the cans before performing the activity.

There are alternatives to using cans: The type of PVC pipe used by plumbers may be substituted for cans in this activity. This pipe is extremely durable. Even in hard soils it can be used over and over by groups performing the activity. PVC pipe that is 10 cm (4 inches) in diameter is readily available in plumbing supply stores. Cut the pipe into 20 cm (8 inch) lengths. Use a flat file to bevel one end of each 20 cm section of pipe, forming a cutting edge.

! Flatten any sharp edges on the inside rims of the cans so students will not cut their fingers. Be sure that all students wear safety goggles while driving the can into the ground. Remind them to be careful to avoid hitting their fingers while hammering the cans into the ground.

In extremely hard-packed soils, you may be unable to drive the can a full 5 cm into the ground. The can tends to collapse before it is deep enough in the soil. In that case, simply sink the can into the soil as far as possible and continue with the next step in the procedure. You need to be aware, however, that the results obtained when the cans are driven different distances into the soil will not be directly comparable.

This activity should not harm the soil in any way; however it is good practice to check with the person(s) responsible for maintaining the grounds before you begin, to be sure that this experiment will not interfere with grass cutting or other maintenance.

If possible, set up this activity in several different types of soil—for example, in a well-tended flower bed, on a lawn, on a packed area of playground, or in a sandbox. The differences in absorption rates for these different types of soils should be striking.

One-liter plastic soft drink containers are handy for transporting water to the area of soil being tested.

◆Suggestions for further study

You may wish to repeat this experiment in the classroom with the following changes. Place the can in a tub with a screen on the bottom of the can. Fill can to the 5 cm mark with loose soil. How much does the soil settle when you pour the water in? Fill with packed soil to the 5 cm mark and repeat the questions.

Students can test soil types in the classroom using glass quart jars and a variety of soils—clay, sand, loam, plant material (grass clippings), and mixes of these types. Fill each jar with a different soil type. When all jars are ready, have students pour a volume of water into all the jars at once, then time the rates of absorption.

◆Answers

The following sample answers are based on data collected from a grass-covered lawn. Your answers will depend on the type of soil tested.

3. Location: The test was performed on a lawn.

Plant material present: grass, clover.

Soil condition: The soil was dry. There had been no rain for eight days.

6. Starting height of water = 11 cm.

Note: The can was placed in a sandbox and these measurements were repeated. The starting height of the water was again 11 cm. However, the sand absorbed the entire liter of water in less than 5 minutes. Some beach and desert soils absorb water at comparable rates.

8. Answers will vary. The soil we tested can absorb about 10 cm of water per hour. From these observations we can predict that it probably would absorb 5 cm of rain striking the ground during a slow, steady rainstorm that lasts 10 hours.

It might be able to absorb 5 cm of rain striking the ground during a heavy thunderstorm that lasts one half hour, but that is almost equal to its maximum rate of absorption.

9. If the soil were located on a steep hillside, the runoff from the rainstorm would be much faster. Most of the slow rain might be absorbed, but the rain from the heavy thunderstorm would probably run off the hillside before it could be absorbed.

10. If the area had not received any rain for a week before the experiment, the rate of absorption would probably increase. On the other hand, if the area had been experiencing a prolonged drought, the soil surface might be so dry and tightly packed that the water would take more time to begin to soak in. (This is similar to the problem of using a completely dry sponge to pick up a small spill of water.)

If it had rained heavily the night before the experiment, the soil would still be nearly saturated with moisture and the rate of absorption would decrease.

Sample data

Centimeters of water absorbed by soil

Time *(minutes)*	Distance between starting-height marker and the surface of the water
5	*1.8* cm
10	*4.2* cm
30	*6.0* cm
60	*10.0* cm

ACTIVITY 3 WORKSHEET

Is It Full Now?

Materials

Each group will need

- 7 small, transparent containers—clear plastic 250 ml (5 ounce) cups are ideal
- marbles or pebbles
- fine, dry sand—enough to fill three of the cups
- a water source
- an eyedropper

◆Background

In this activity you will use marbles and sand to make model soils. You will use these "soils" to investigate the relationship between particle size, pore size, and the water-holding capacity of soil.

◆Objective

To demonstrate the presence of pore spaces in sediments and rocks

◆Procedure

Before class, your teacher dried the sand by spreading it on a cookie sheet, placing it in an oven set at 180° C (about 350° F) for 15–30 minutes, and stirring it several times while it was heating.

1. Fill the 7 cups with the materials listed below. You should fill each cup to the same height—a point just a little below the rim.

- Fill 2 cups with marbles (or pebbles).
- Fill 2 cups with fine, dry sand.
- Fill 3 cups with water.

Vocabulary

- **Pore spaces:** Open areas, or spaces, in soil, sediments, and rocks that are filled by air or water. If pores are connected, water can flow through the material (the soil, sediment, or rock).

All seven cups are now "full." However, if you look closely, you can see that there are spaces remaining between the marbles and grains of sand.

2. Which type of "full" cup do you think has the greater total volume of "empty space" (pores) between its particles—the cup of sand or the cup of marbles?

---

3. Check your answer to question 2 as follows: Pour water from one of your full water cups into a "full" cup of sand until the sand is completely wet. Pour slowly; several minutes may be required for the water to soak all the way to the bottom of the cup. When the sand is saturated, the water should be level with the top layer of sand. If you pour in too much water, use an eyedropper to put it back into the container of water.

The cup is now "full" of sand *and* water; the water now fills the pores between the grains of sand. Estimate how much water was needed to fill the pores in the sand (1/10 of the container of water, 1/3, 1/2, etc.).

Estimated amount of water needed to fill the pores in the sand:

____

About how much "empty space" (pores between the grains of sand) was there in the container that was "full" of dry sand?

____

Set the containers of wet sand and water aside so that you can compare them to other cups later.

4. Now pour water from the second full water cup into a "full" cup of marbles. Add water until the liquid is even with the tops of the marbles.

Estimate how much of the cup of water was required to fill the pores between the marbles. (1/10 of the container, 1/3, 1/2, etc.).

Estimated amount of water needed to fill the pores between the marbles:

____

5. Compare the amount of water needed to fill the pores between the marbles with the amount of water needed to fill the pores between the sand. On the basis of your observations, place a check mark beside the following statement that best describes the amount of pore space between the grains of sand and the marbles:

a) There is more pore space ("empty space") between the grains of sand than between the marbles.

b) There is more pore space (empty space) between the marbles than between the grains of sand.

c) The pore space (empty space) between the marbles is about the same as the pore space between the grains of sand.

When you have completed this comparison, set all of the cups aside for later use.

6. Suppose that you fill the pores between dry marbles with sand rather than with water. Will you need the same volume, a smaller volume, or a greater volume of dry sand to fill the "empty space"? Why do you think this is the case?

____

____

7. Check your answer to question 6 by pouring sand from your remaining "full" cup of dry sand into the remaining "full" cup of marbles. Use the following procedure:

Tap and gently shake the marble cup while you pour sand onto the marbles. The sand will fall into the spaces between the marbles and fill the "pores" all the way to the bottom of the cup.

Stop pouring when the sand has filled all of the pore space and is level with the tops of the marbles. The cup is now "full" with a mixture of sand and marbles. Estimate how much sand was required to fill the pores between the marbles (1/10 of the "full" cup of sand, 1/4, 1/2, etc.).

Estimated amount of sand needed to fill the pores between the marbles:

____

Were you able to fit more sand or more water into the pores between the marbles? ________

8. Find out how much "empty space" remains in this cup "full" of marbles and sand by pouring water from your last full water cup on top of the sand and marbles.

Continue pouring water onto the marbles and sand until all the sand is wet and the water is level with the tops of the marbles.

Estimate how much of the cup of water was required to fill the pores between the marbles and sand. (1/10 cup, 1/3, etc.).

Estimated amount of water needed to fill the pores: ________

9. Which of the three cups had the least amount of pore space (empty space) between its particles: (a) the "full" cup of sand, (b) the "full" cup of marbles, or (c) the "full" cup of sand and marbles?

GUIDE TO ACTIVITY 3

Is It Full Now?

◆What is happening?

All soil contains pores. **Porosity**, the ratio of open pore space to the total volume of a soil sample, is one of the factors that determines how much water a particular soil can hold and how fast water can move through that soil.

The sizes of the particles in soil determine the size and number of pores in soil. The size of soil particles also affects the ease of plowing the soil, what crops can be grown, the efficiency of certain fertilizers, and the ability of soil to store water.

The marbles and sand demonstrate that the pore space is larger in coarse soils. In actual coarse soils, these spaces may be partially filled with smaller particles, producing a less porous soil. When small particles fill the large pore spaces, there is less "empty space" remaining for water to fill. However, soil containing a mixture of large and small particles *retains* its water more efficiently than coarse soils, because the small particles provide more surface area for the water to adhere to. Coarse soils dry out much faster than do fine soils. You may have noticed, for example, how quickly beach sand dries after a rainfall.

Soil scientists classify soils according to their particle size. Clay particles are the smallest (less than 0.004 mm in diameter), followed by silt (up to 0.06 millimeters in diameter), then sand (up to 2.0 mm). Particles larger than sand are classified as gravel, and range in size from very fine pebbles to large boulders.

Vocabulary

- **Hypothesis:** An informed guess.

◆Time management

One class period (40–60 minutes) should be enough time to complete the actual class procedure and cleanup. A good part of another period could be used to discuss and explain the results.

You may want to plan for discussion time between the steps of this activity, especially between steps 6 and 7.

◆Preparation

You can buy sand for this activity at most home and garden supply stores. Any reasonably clean, fine sand may be used. The type of sand sold in large bags for use in children's sandboxes works very well.

Dry the sand by spreading it on a cookie sheet and placing it in an oven set at 180°C (350°F) for 15–30 minutes. Stir it several times while it is heating. Sand dried in this manner may be stored in plastic bags for several weeks.

This activity contains frequent references to the "empty space" between soil particles. Remind your students that this empty space contains air. The water poured into the empty space is not filling a vacuum—it is displacing the air that surrounds the particles of sand or the marbles.

You may wish to perform this activity as a demonstration for your classes. If so, we recommend that you place the sand, marbles, and water in larger containers (such as 500 ml beakers) so that the results are easier for everyone to see. The results will be more convincing to younger students if all of the containers are exactly the same size and shape.

You can use almost any type of container for this activity, as long as you have a sufficient number of containers of the same size. Small (150 ml) beakers will work very well.

! If using beakers or any glass containers, be sure to put the marbles in them *carefully* so that the containers do not crack.

Using marbles for this activity makes it easy for students to see the "empty space" between them. For a more realistic demonstration, marble-sized pebbles may be substituted for marbles. The container filled with pebbles may have a different proportion of empty space (pores) than the container of marbles. It will also be somewhat harder to fill the pores between pebbles with dry sand; you will have to shake the container of pebbles vigorously to ensure that the sand reaches its bottom.

◆Suggestions for further study

You may require advanced students to perform this activity as a more quantitative exercise by having them measure (in milliliters) the amount of water required to completely fill the cups containing the marbles, sand, and sand and marble mixture. These measurements will show that the "full" container of sand is actually about 60% "empty space", the cup of marbles is about 50% empty space, and the sand-and-marble mixture is about 20% empty space.

Exact quantitative measurements are not essential, however. The point of this exercise is to show that there are pores between soil particles. Sometimes the pores are large and easy to see (as with the marbles), and sometimes the pores are not obvious until you fill them with a liquid (as is the case with the cup of sand). Estimating what fraction of a cup of water is required to fill the "empty space" between the marbles or sand will demonstrate to students that there is a lot of pore space in soil.

◆Answers

2. Answers will vary, but many students will decide that the large, obvious openings between the marbles constitute more "empty space" (that is filled by air) than the tiny pores around the tightly packed sand granules.

3. About 1/2 of the container of water will be required to fill in the pore spaces between the grains of sand.

About half of the "full" cup of sand is "empty space."

4. About 1/2 of the container of water will be required to fill in the pore spaces between the marbles.

5. The correct answer for the model soil we tested was (c). The "full" cup of marbles and the "full" cup of sand each contain about the same amount of "empty space," so are about equally "full."

6. This question asks students to state a hypothesis. Their answers will vary—do not emphasize any right answer. Students will test their hypotheses in subsequent steps and decide for themselves whether or not their guesses were correct.

Before allowing the class to go on to the next step, you may wish to encourage students to discuss their reasons for giving different answers.

7. Less than 1/2 of the container of sand will be required to fill in the pore spaces between the marbles.

More water will fit into the pores between the marbles.

8. About 1/5 of the container of water will be required to fill in the pore spaces between the marbles–sand mixture.

9. The cup of sand and marbles had the least amount of "empty space" between its particles. The full cup of sand and the full cup of marbles each had about the same amount of "empty space" between their particles.

ACTIVITY 4 WORKSHEET

How Much Water Can Different Soils Hold?

◆Background

Soil consists of small particles of broken rock mixed with organic material (materials produced by living things). Both the size of its particles and the amount of organic material it contains affect a soil's water-holding ability.

◆Procedure

1. If your teacher or workshop leader has not done so already, a day or two before beginning the activity, dry the soil samples as follows:

First, break up any large clods in the soil. Spread the soil samples out on separate cookie sheets and place them in an oven set at about 95° C (200° F) for 30 minutes. Allow the samples to cool to room temperature. When they are cool, transfer the samples to two separate paper bags. Write "rich" on the bag containing the dark soil and "poor" on the bag containing the other sample.

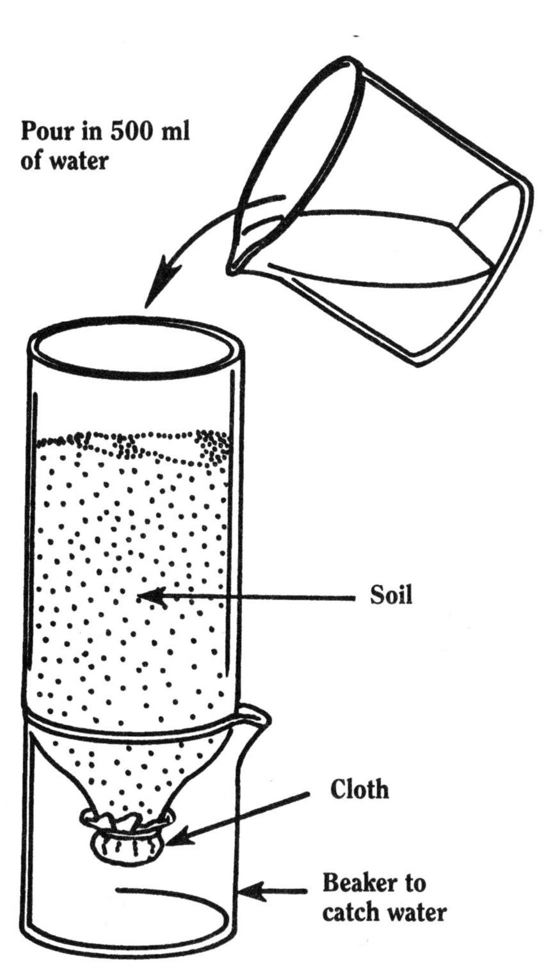

2. Make two cylinders by cutting the bottoms off the two soft drink containers. Place a square of cloth over the mouth of each bottle, and secure it with a rubber band.

3. Mark one bottle "rich soil" and fill it about 3/4 full with the rich soil sample. Mark the other bottle "poor soil" and fill it 3/4 full with the second soil. (Make sure that both bottles contain the same amount of soil.) Stand each bottle vertically in a beaker.

4. Label one of the two remaining beakers "water for rich soil" and the other beaker as "water for poor soil." Pour 500 ml of water into each of these beakers.

Record the time in the data table at the top of the next page. Begin pouring 500 ml of water into each soil-containing bottle.

As the water sinks into the soil, add more water to each soil sample using *only the water remaining in its own water beaker.* Record your observations in the following data table:

Materials

Your teacher will give you instructions on where and how to collect soil samples for this activity.

Each group performing this activity will need about 1 L of each of two types of soil—enough of each soil type to fill a large soft drink bottle about three-quarters full.

(1) One soil sample should be "rich" (containing organic material)—black or dark brown in color, crumbling easily, and free of clods.

(2) The second soil sample should be "poor" or "worn out" (lacking organic matter)—lighter colored, hard, and cloddy.

Both soil samples must be dried before performing the activity; the drying procedure is given at left.

- 2 dry soil samples—one that is "rich" and one that is "poor"
- 2 clear plastic disposable soft drink bottles (1 L size)
- 4 beakers (500 ml beakers or similar size jars may be used)
- scissors or knife to cut drink bottles
- 2 small (10 cm x 10 cm) squares of cloth
- 2 rubber bands
- a 500 ml graduated cylinder
- a waterproof marker
- a watch or clock with a second hand

Water absorption data

	Rich soil	Poor soil
Time when water is added	___________	___________
Time when dripping begins	___________	___________
Time when dripping stops	___________	___________

5. How many minutes did it take for the water to begin dripping into the beaker from

a) the rich soil? __________ minutes

b) the poor soil? __________ minutes

6. Twenty minutes after you began adding water to the soil, determine the amount of water that was *not* absorbed as follows:

a) For your rich soil sample, pour any of the original 500 ml of water that remains in the beaker into the graduated cylinder. Then, carefully pour the water left standing on top of the soil into the same graduated cylinder. Try to avoid disturbing the soil as you pour the water off of its surface. Now, determine the volume of water in the cylinder.
In the "Unabsorbed Water Data" table, record this volume of water under the heading "Volume of water not entering the soil."
b) Determine the volume of the water that passed *through* the rich soil sample. This is the water that dripped into the lower beaker. Enter this volume in the table below.
c) Calculate the amount of water that was not absorbed by the rich soil sample by adding the volume of water that did not enter the soil to the volume of water that passed through the soil. Enter this sum in the data table.
d) Repeat steps a, b, and c for the "poor" soil sample.

Unabsorbed water data

After 20 minutes:	Rich soil	Poor soil
Volume of water that did not enter the soil	______ ml	______ ml
Volume of water that passed through the soil	______ ml	______ ml
Total unabsorbed water	______ ml	______ ml

7. Determine how much water the rich soil and the poor soil samples *absorbed* by subtracting the volume of unabsorbed water from 500 ml (the starting water volume) for each sample.

Water absorbed by soil sample = 500 ml - Total unabsorbed water

Water absorbed by "poor" soil = 500 ml - ______ ml = ______ ml

Water absorbed by "rich" soil = 500 ml - ______ ml = ______ ml

8. On the basis of your results, decide which of these soils you think would be best suited for growing crops. Explain how you reached your decision.

____

____

____

9. Why do you think that the soil samples absorbed different amounts of water?

____

____

____

GUIDE TO ACTIVITY 4

How Much Water Can Different Soils Hold?

◆What is happening?

Organic material in the soil (humus) acts something like a sponge, taking up water that enters the soil. Organic material also helps to keep the soil's pores open. As a result, soil rich in humus absorbs moisture quickly. Quick water absorption means that there is little water left on the surface to become runoff, so humus-rich fields are less likely to be damaged by erosion than fields that lack humus.

Humus-rich soils are preferred by farmers and gardeners, because the high water-holding ability of rich soil provides more moisture for plant growth. The more moisture that enters the soil, the more productive a field is likely to be. In dry areas, humus-rich soils require less irrigation to produce the same crop. Since irrigation costs the farmer money, fields that retain their soil moisture are more profitable.

If your soil samples were in a farmer's field rather than in plastic containers, the water that passed through the soil would not drip out into a beaker—it would continue sinking deeper into the subsoil toward the water table. Although deep-lying soil moisture may not be directly accessible to plants growing on the surface, it does recharge groundwater supplies. Therefore, having rainwater soak into the subsoil is preferable to allowing much of it to become runoff.

◆Time management

Not counting the time to gather the materials and dry the soil, one class period should be enough to complete this activity. You can assign a reading activity for the 20 minutes you are waiting to collect the second set of data.

◆Preparation

Having students collect the soil used for this activity encourages them to think about differences in soils found in a variety of locations. However, you must give your students clear guidelines about where they may or may not collect soil for this activity. If you live in a densely-populated area, it is a good idea to have students get written permission from the landowner before removing any soil. (A professional soil scientist would certainly get permission before conducting a study on private property!)

Collect (or have students collect) about 1 L of each of two types of soil for each group performing this activity—enough of each soil type to fill a 1 L soft drink bottle about three-quarters full. One soil sample should be "rich"—containing organic material, black or dark brown in color, crumbling easily, and free of clods. Commercial potting soil will work. The second soil sample should be "poor" or "worn out"—lacking organic matter, lighter-colored, hard, and cloddy.

On your own, or as a class, a day or two before performing the activity, remove any excess moisture from the soil samples using the following procedure:

First, break up any large clods in the soil. Spread the soil samples out on separate cookie sheets and place them in an oven set at about 95° C (200° F) for 30 minutes. Allow the samples to cool to room temperature.

(Note: Heating the soil until it is "dry as dust" is not necessary. The object of this step is to insure that both soil samples contain about the same amount of moisture before you place them in the cylinders.)

When the soil samples are cool, transfer them to two separate paper bags. Write "rich" on the bag containing the dark soil and "poor" on the bag containing the other sample.

Make "cylinders" by cutting the bottoms off the 1 L soft drink containers. Place a square of cloth over the mouth of each bottle and secure it with a rubber band.

Plastic soft drink bottles make ideal cylinders for this type of activity. We have found that the 1 L bottles are most convenient to use, but either 2 L or 450 ml (16 oz) bottles will also work; you will, however, need to adjust the volume of water applied to the soil samples if you use 2 L or 450 ml bottles.

◆Suggestions for further study

Study the two soil samples under a microscope. With a soil identification kit as a guide, identify the components that make up the soils. What are the origins of the components? If students have had any introduction to the geologic history of their area, how do the components of the soil they see under the microscope reinforce their knowledge?

As you travel around your state, collect soil samples in quart jars to use as references with the activities of this module. It will become apparent why some areas are agricultural and others are not.

◆Answers

Note: The following data and answers are intended to serve only as general guidelines. The actual results that you obtain will be determined by the type of soil that you test.

4.

Sample data

Water absorption data

	Rich soil	**Poor soil**
Time when water is added	*0 min*	*0 min*
Time when dripping begins	*1 min 25 sec*	*18 min*
Time when dripping stops	*takes longer than 20 min*	*takes longer than 20 min*

5. Answers will vary depending on the type of soil that you are using. In many cases, the rich soil will drip first. However, if the poor soil sample is mostly sand, it may drip first. A sandy soil sample that we tested absorbed all 500 ml in less than six minutes.

Sample data

Unabsorbed water data

After 20 minutes:	**Rich soil**	**Poor soil**
Volume of water that did not enter the soil	*75* ml	*225* ml
Volume of water that passed through the soil	*150* ml	*10* ml
Total unabsorbed water	*225* ml	*235* ml

7 & 8. Answers will vary. For the sample data shown above, the soils absorbed almost the same amount of water. This does *not* mean, however, that these two soil samples were "equivalent."

The water entered the rich soil much faster; all but about 15% of the 500 ml entered the soil within 20 minutes. A large volume (in this case, 150 ml) dripped out the bottom of the soil sample. In a farmer's field, the water that passes through the top layers of soil is absorbed by the lower layers of soil, increasing the soil's moisture content. Some absorbed water sinks to the water table.

Although the poor or worn-out soil eventually absorbed about the same amount of water as the rich soil sample, it did so very slowly. Almost half of the 500 ml of water was still on the surface of the soil after 20 minutes. In most farmers' fields, there is enough slope to the land's surface that water runs off rather than sitting on the surface of the ground. If the soil of a farm absorbed water as poorly as this sample did, most of the water would become runoff. Runoff is not available for plants, and causes erosion.

9. The humus (decayed plant and animal material) in the soil probably accounts for much of the different patterns of water absorption. Humus absorbs water efficiently and helps to create channels and air pockets in the soil.

GUIDE TO ACTIVITY 5

Who is Responsible for Preserving Anyville's Water Supply?

◆What is happening?

This role-playing activity is designed to show that there can be many different but quite reasonable opinions regarding the regulation of underground water resources. There is no right or wrong solution to Anyville's problem. Whatever decisions are made regarding water usage will have an effect on people's jobs, finances, and quality of life.

◆Preparation

Duplicate "Anyville, USA: A Profile." Make enough copies so that each student can read this information before hearing the presentations of the various special interest groups that will testify at the Anyville town meeting.

Duplicate the "Briefs" of each special interest group. Only those students who will represent a particular special interest group during the town meeting should receive a copy of the group's brief.

Have all students read "Anyville, USA: A Profile." This gives general information about Anyville and its water problem. It also lists the special interest groups testifying at the town meeting being held to discuss issues concerning water usage.

Give students who will play the role of special interest group members copies of their interest group's Brief. These students must prepare to present their side of the water issue at the town meeting; they must also be able to defend their position against the objections and arguments of the other citizens of Anyville.

Class members who are not presenting special interest group positions can serve as jurors. They will decide, after listening to all the arguments, what action (if any) should be taken to solve Anyville's water problem. Encourage jurors to take an active role when various special interest group positions are being presented—allow jurors who want more proof or have questions about a particular position to share their thoughts with the class.

You can play the role of moderator or judge for the presentations. Try to be impartial as the debate develops—there is no right or wrong solution to Anyville's problem. Your personal "correct solution" to Anyville's problem may be quite different from your students' solution.

Remind your students that whatever decisions are made regarding water usage will affect jobs, finances, and quality of life. Different groups have different priorities on water issues; each group, if given the opportunity, will allocate water resources in the manner that helps it achieve its goals.

◆Suggestions for further study

Invite a professional soil scientist from the Soil Conservation Service or a consulting firm to talk to the class about how he or she would address the controversy. Be sure to go over the activity with the professional before class. There can be several "right" ways to solve problems. If students are interested, another professional soil scientist could be invited to present another point of view.

For all students

Anyville, USA: A Profile

Anyville has a population of 10,000 people, of whom 7000 are employed outside the home. There is a municipal water supply system that draws water from six deep wells. This system supplies water to houses in the older section of town. New housing developments on the north and east sides of town do not receive water from the municipal supply; homeowners there must drill individual wells to supply water for household use.

Anyville's water problem was first noticed by the residents of the two new housing developments. In the five years since the houses were built, the water table has dropped and many of the homeowners in both developments were forced to drill deeper wells to meet their families' needs. During the same time period, the town drilled two additional wells in order to meet rising demands for water, and to offset the declining yield of its four existing wells.

Farms surround the town, and 50% of the town's economy is related to agriculture—this includes banking and legal services; selling farm implements, seeds, and fertilizer; and marketing farm products.

A fabric production plant is the town's major nonagricultural employer. The plant employs 500 full-time and 300 part-time workers—over 10% of Anyville's work force. Originally a cotton mill, the plant was redesigned in the 1960s to dye and process synthetic fibers. The plant has its own large well and treatment facility to provide the water required for the manufacturing process. The plant's wastewater is pumped into holding ponds on the site of the plant; sludge from these ponds is shipped periodically to a chemical waste landfill 150 miles away.

The story thus far

The town of Anyville, USA, has a serious problem: the supply of groundwater on which the town depends is running low, and there is evidence that the remaining groundwater is becoming polluted.

Maintaining the water supply is vital to the citizens of Anyville. Without an adequate water supply, industries may leave the town, causing unemployment; farmers in the surrounding areas may go out of business if they cannot irrigate their crops; homeowners might be forced to buy bottled water for drinking and washing—the list of consequences goes on and on.

The mayor of Anyville has called a town meeting to discuss possible solutions to the water problems. Various special interest groups who are concerned with regulating water use will present their views about the water issue.

The following are the special interest groups testifying at the hearing on Anyville's water use:

- Association of Concerned Homeowners of East and North Anyville
- XYZ Fabric Manufacturing, Inc.
- Anyville Town Manager (on behalf of Anyville Town Council)
- Anyville Merchants' Association
- Anyville Regional Farmers' Association

If the citizens of Anyville cannot come to an agreement about preserving the water supply, any lawsuits that result will be decided in state court according to the "reasonable use" doctrine. This doctrine states that a landowner's right to use groundwater can be limited by the court when the available supply is not sufficient to meet all the demands placed on the supply.

Brief: Association of Concerned Homeowners of East and North Anyville

For students assigned to this role only

◆Background on the Homeowners' Association

This group represents 200 owners of homes in two developments that were built five years ago. These homes have an average value of $150,000. Since the Anyville Waterworks does not supply water to the area in which these homes were built, each homeowner has a well to supply all the water necessary for household use.

Within the last two years, 27 homeowners have experienced such severe water shortages that they have had to dig new wells at an average cost of $2500 per well. A total of 56 other homeowners have experienced reduced output from their wells, but they solved the problem by using water conservation measures.

Six homeowners living in the eastern development noted a more serious problem: The water from their wells had developed a funny taste. When they had the water tested, they discovered that it contained chemical contamination: low levels of solvents and synthetically produced chemicals. The water was pronounced unfit for human consumption, and the homeowners were forced to buy bottled water for washing and drinking. This problem has not cleared up.

◆Homeowners' position

The homeowners suspect that the XYZ Fabric Manufacturing plant, despite its claims to the contrary, is responsible for overdrawing the underground water supply and causing the water shortage. Further, they suspect that the plant may be contaminating the groundwater with the solvents used in the dyeing process for fabrics. They fear that they will be unable to sell their houses, which in most cases represent their life savings, if water is not available.

◆Homeowners' solution

Make XYZ reduce its water use and install a new waste treatment facility on the plant grounds. Force XYZ to pay for drilling a new well for any home that needs one, and pay damages to owners whose wells are contaminated with organic chemicals. Alternatively, the homeowners want Anyville to extend the municipal water lines to their houses.

For students assigned to this role only

Brief: XYZ Fabric Manufacturing, Inc.

◆Background on this company

XYZ is Anyville's largest employer. XYZ directly employs over 10% of all Anyville workers; their earnings are spent mainly on the goods and services provided by the merchants in Anyville. Through direct payments of business taxes and indirect payments in the form of sales and property taxes of its employees, XYZ is by far the largest taxpayer to the town, accounting for almost a quarter of the town's total tax revenues.

Company position: XYZ has always complied with all state and local laws regarding water use and waste disposal. The amount of water it uses is determined by the amount of fabric being produced; in order to reduce water use, the company would have to reduce production and lay off workers.

The company categorically denies that it is contaminating the underground water resources of the area and suggests that the real source of pollution is the leaching of materials from an old dump that Anyville closed ten years ago. The dump received not only domestic refuse, but also the waste from a small chemical plant that went out of business 20 years ago. XYZ suggests that homeowners' wells are failing because of increased farm irrigation, which is a legal use of the underground water supply.

Further, the company states that if any additional water use or waste disposal regulations are imposed on XYZ, it will be forced to consider relocating the plant. If this happens, not only will XYZ's employees be without jobs, but the merchants in Anyville will face economic hardships as well, and the town may have to reduce its services because of the loss of tax revenue from XYZ.

◆XYZ's solution

Anyville should excavate the old dump, remove the hazardous materials, and "seal" the bottom of the dump with a combination of a plastic barrier and special impermeable clay. The company also suggests encouraging local farmers to grow crops that require less irrigation.

Brief: Anyville Town Manager
(on behalf of the Town Council)

For students assigned to this role only

◆Background on Anyville

Because of a decline in the regional farm economy in the last ten years, many of the agriculture-related businesses that previously made up a large part of Anyville's tax base have failed. As a result, Anyville has been forced to lay off police and firemen and cut back on school financing. Any additional expenditures or loss of tax revenue would have serious consequences for the town.

◆Anyville's position

Within the present tax structure, Anyville cannot afford to extend water service to the new housing developments in North and East Anyville. Doing so would necessitate drilling additional town wells, in addition to expanding water treatment and pumping facilities. The total cost would be in excess of one million dollars.

The town does not accept responsibility for polluting the underground water by improper use of the old dump. The landfill facility was built and used in compliance with the state and national codes in existence at that time. Cleaning up the old dump and sealing it against leakage is not feasible, because it would cost in excess of two million dollars.

◆Anyville's solution

Observe the water use patterns and monitor pollution levels for a study period of five years. If, at the end of that time, the pollution and water shortage situations have not improved, form a committee to investigate means of dealing with the problems.

If immediate action is taken, the following will be required: passage of a three million dollar bond issue, coupled with a 50% increase in property taxes and a further reduction in fire and police protection. School budgets will also be reduced, beginning with the elimination of "frills" such as football and basketball teams and other school-supported extracurricular activities.

For students assigned to this role only

Brief: Anyville Merchants' Association

◆Background on the Merchants' Association

This group was founded to help Anyville merchants compete with similar businesses that have been established at a recently built shopping mall that is only 20 minutes away by car. Although their businesses are not directly threatened by the water problem, they know that any loss of jobs as a result of closing the XYZ plant would force many of them out of business. An increase in taxes to pay for water projects would make them less able to compete successfully with the shopping mall and might also put some people out of business.

◆Merchants' position

Is there really a problem? Everybody knows that some wells run dry when the rain is scarce; the last three years have been among the driest in Anyville in the last 40 years. When the rain gets back to normal, the homeowners will not have to worry about water shortages; in the meantime, that is just a risk that a person buying a home has to take.

As to the contaminated wells: Isn't it possible that the water was contaminated from some source other than the landfill or the XYZ plant? It is quite possible that the homeowners themselves contaminated the wells by improper use of pesticides or lawn treatment chemicals. After all, only six wells were affected out of several hundred; raising taxes to clean up the landfill or possibly forcing the town's biggest industry to leave seems like a very drastic action, in light of the small amount of inconvenience that those few homeowners face.

◆Merchants' solution

Why not wait for a few years and see what happens before spending all of that tax money?

Brief: Anyville Regional Farmers' Association

For students assigned to this role only

◆Background on the Farmers' Association

This group was founded five years ago as a cooperative effort to help family farmers cope with declining farm prices. The drought of the last three years has put even more farmers in danger of losing their land and going out of business. Only those farmers having the capital necessary to buy equipment and drill wells to irrigate their farms extensively have turned a profit during those years; many others have gone bankrupt and lost their land.

◆Farmers' position

Farmers are as concerned as any other group about the condition and availability of groundwater supplies. Farmers not only use groundwater to irrigate crops, they drink groundwater and feed it to their livestock, too. The deeper the water is in the ground, the more it costs to drill wells and to pump the water out; the more money spent on irrigation, the less profit farmers make. The local farmers cannot irrigate less, or many of them will certainly go broke.

The farmers believe that the real "water thief" that is stealing the supplies of Anyville is the weather. The longer the drought persists, the greater will be the demand on groundwater for agricultural use. Increased use of groundwater lowers the water table and causes wells to run dry. As the water table declines, possible sources of pollution that had never before caused problems may begin to contaminate wells.

◆Farmers' solution

There is no solution to a drought but a change in the weather—all we can do is wait for the rain to come.

MODULE 2

Reshaping the Surface of the Earth

◆Introduction

• If all the topsoil that the Mississippi River carries past Memphis, Tennessee, during a single hour could be collected, there would be enough soil to form 40 acres of good cropland.

• In the Mississippi delta, land near oil wells is sinking as the oil is pumped out. If the present rate of land sinking continues, the farms and towns in some of these areas may soon be covered by the waters of the Gulf of Mexico.

• Most Americans have never seen a **glacier**, but in North America more water is stored as ice or snow in glaciers than in all the lakes and streams on the continent.

In the brief span of a human life, the surface of the Earth seems unchanging. Nothing could be further from the truth. Every day, solid rock is being broken into pebbles, sand, and smaller sediments by the geological processes of weathering, including the freezing and thawing of water. Over time, streams will carry entire mountain ranges into the ocean basins.

Before we look at how water reshapes the surface of Earth, it will be useful to define a few terms. **Sediment** refers to Earth material in a variety of sizes, ranging from large boulders to tiny particles of clay. **Soils** are composed of sediments combined with organic matter (decayed plant and animal material).

Scientists use "grain-size scales" to label sediments. For example, American geologists refer to particles smaller than 0.004 mm in diameter as clay size. Next larger in size are particles called silt (up to 0.05 mm), followed by sand (up to 2.0 mm). All particles larger than sand are classified as gravel. Gravel ranges in size from pebbles (up to 64 mm), to cobbles (up to 256 mm), to boulders. Any sediment larger than 256 mm in diameter (about 10 inches, or a bit larger than a basketball) is referred to as a boulder.

Vocabulary

• **Glacier:** A large mass of ice formed on land by the compacting and recrystallization of snow. Glaciers survive from year to year, and creep downslope or outward due to the stress of their own weight.

• **Sediments:** Fragments of material produced by weathering and erosion of rocks. Sediments may remain on the Earth's surface as soil, or may be transported and deposited in other locations by wind, streams, and other erosional agents.

• **Soil:** Sediment on or near the Earth's surface that is formed by the chemical and physical weathering of rocks as well as the decay of living matter. Soil is sediment capable of supporting the growth of land plants.

Grain-size scale used by American geologists*

Grade limits (minimum diameter)		Grade name
mm	*inches*	
256	10.1	boulders
64	2.5	cobbles
2	0.08	pebbles
0.05	0.02	sand
0.004	0.002	silt
smaller than silt		clay

A single lifetime is too short to observe most major effects of water acting on the land. However, scientists are learning to detect the changes that have already taken place, and are studying the forces that reshape the surface of the Earth. In this module, you will view some of the evidence that great changes have taken place on Earth, and work with experimental models that simulate the means by which the face of the Earth is constantly being changed.

Moving water is a major factor in sculpting our landscape. Streams produce valleys, flood plains, and deltas through the processes of erosion and deposition of sediments. The presence of flowing water on Earth, and the geological features produced on the planet's surface by flowing water, help distinguish Earth from other celestial bodies.

In this module we will study how water flowing down a stream moves solid materials. Rivers, brooks, creeks, rills, kills, and tributaries all are names for different types of streams. All of these have several things in common: (1) they consist of liquid water (2) the water flows through a channel, and (3) their water source is above sea level.

Not all moving water is liquid, however. Glaciers, large masses of snow and ice that never completely melt over a period of many years, contain most of Earth's fresh surface water. Advancing and retreating glaciers help shape mountains and valleys, move countless millions of metric tons of soil, and are the sources of many of the world's rivers.

In this module you will explore some of the characteristics of frozen water and review the evidence for the last great ice age that reshaped much of the northern part of our continent.

*Source: Data Sheet 17.1, American Geological Institute.

SUMMARY OF MODULE 2

Reshaping the Surface of the Earth

◆Instructional objectives

After completing the activities and readings for Module 2, you should be able to

- explain how water contributes to the process of erosion [Activities 6, 10, and 11]
- build a model of a stream [Activity 7]
- measure the rate of flow of a stream [Activity 8]
- describe the factors that affect the rate of flow of streams [Activity 9]
- demonstrate four ways that streams carry sediments [Activity 10]
- show how the abrasive action of glaciers can change the surfaces that they flow across [Activity 11]

◆Preparation

Study the following readings for Module 2:

Reading 5: Flowing Water Reshapes the Earth

Reading 6: Rivers of Ice: Glaciers

Reading 7: Danger! Rising Water!

Reading 8: Scientific Literacy for All

Reading 9: Water and the Jobs of Your Lifetime

◆Activities

This module includes the following activities:

Activity 6: Film: "Erosion: Leveling the Land"

Activity 7: Building a Model of a Stream

Activity 8: Calculating the Rate of Flow of a Stream

Activity 9: What Factors Affect the Speed of a Stream?

Activity 10: How Can Streams Move Mountains?

Activity 11: How Ice Shapes the Land—A Glacier on the Rocks

ACTIVITY 6: FILM

Encyclopaedia Britannica Film: "Erosion: Leveling the Land"

Vocabulary

- **Deposition:** The process of dropping or getting rid of sediments by an erosional agent such as a river or glacier; also called sedimentation.
- **Erosion:** The processes (including soil erosion) of picking up sediments, moving sediments, shaping sediments, and depositing sediments by various agents. Erosion plays a role in creating Earth's surface features—the landscape. Erosional agents include streams, glaciers, wind, and gravity.

◆Background

This film examines the effects of geological processes such as weathering, erosion, and deposition of rock materials. It shows laboratory studies of the effects of water on rock, and also uses techniques such as time-lapse photography to show how streams of flowing water can undercut a bank, causing it to collapse.

◆Objective

To explain how water contributes to the process of erosion

◆Time management

The film's running time is 14 minutes. One class period should be allotted for an introductory lecture, viewing, and discussion of the film.

◆Summary of film content

Weathering causes gradual disintegration and crumbling of rock on Earth's surface. In a wet and changeable climate, this process proceeds more quickly than in a dry climate. Why? The answer can be found by carefully examining some of the different ways in which rocks weather:

- In rainy regions, rocks are repeatedly wetted and dried, and the minerals of which they are made decompose in somewhat the same way an iron nail rusts in water.
- Some rocks, like limestone, are weather-resistant in dry climates but are readily dissolved and washed away by rainwater.
- In changeable climates, repeated freezing and thawing of water causes cracked rock to break apart.

Once rock is broken apart by weathering, gravity pulls the sediment downhill. This downward movement can be spectacular, as in avalanches and landslides, but it is occurring everywhere in a more subtle fashion—by the creeping and slumping of soil and weathered rock.

When weathered rock, or rock debris, reaches a stream bank, it begins its trip toward the sea. Along the way, some of this material may be temporarily trapped in an inland valley or lake, but most of it will eventually be carried to the seas and will settle on the ocean floor. The removal of rock debris from high ground by streams, rivers, glaciers, and other natural agents such as wind is called **erosion**. The accumulation of sediment on low ground and ultimately on the seafloor is called **deposition**. The net effect of erosion and deposition is to level the land.

Source: Encyclopaedia Britannica Educational Corporation study guide for the film.

ACTIVITY 7 WORKSHEET

Building a Model of a Stream

◆Background

In this activity, you will set up a model stream in a piece of plastic rain gutter. You will vary the slope of the streambed and vary the amount of water flowing into the stream from its source.

◆Procedure

1. Place the trough on a table so that the 1-cm reference line (1 cm from the end of your trough) is aligned with the edge of the table and a short section of trough extends beyond the edge of the table.

Place supports under the trough so that its raised end is 10 cm above the table surface.

Stack supports for the jug near the raised end of the trough. The base of the jug should be about 14 cm above the surface of the table (4 cm above the raised end of the trough).

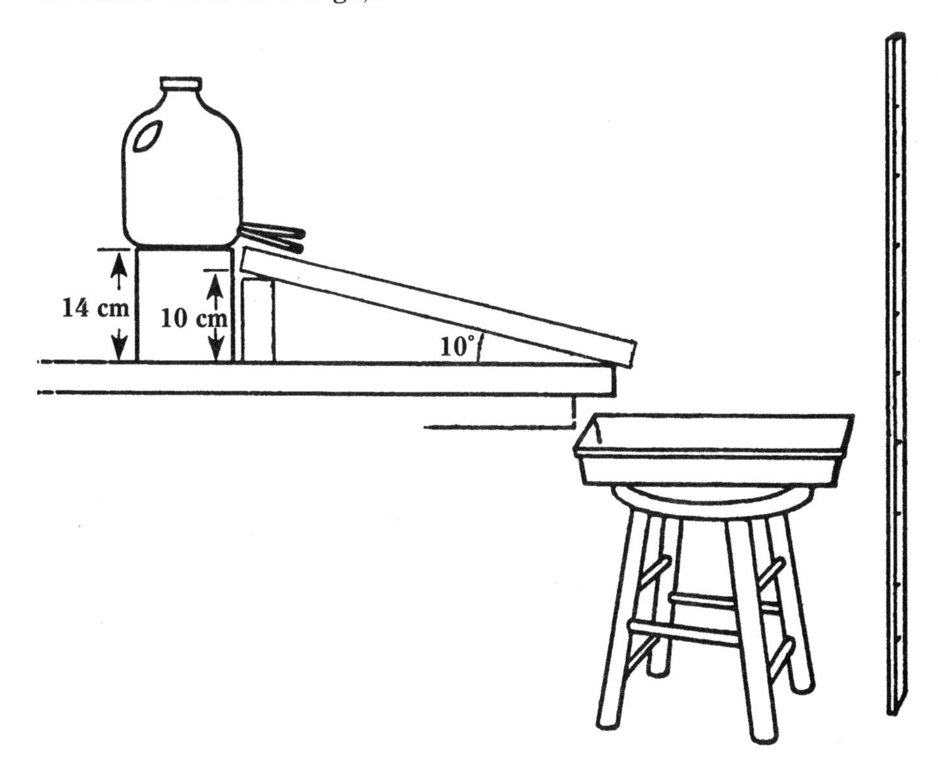

2. Place the collecting pan underneath the lower end of the trough, as close to the overhanging end as possible. (You may wish to set the collecting pan on a stool or chair if you are using a high table.)

3. Be sure that the pencils stuck in the holes in the bottom of the jug are securely in place. Fill the jug with water (preferably over a sink).

4. Set your full jug on the support with its holes facing the trough. Remove one pencil. The water should flow down the trough and into the collecting pan. If there is much spillage, adjust the position of the jug, trough, and collecting pan and try again.

Materials

Each group will need

- a stream trough
- a 3.8 L (1 gallon) milk jug with two pencils stuck into holes near its base
- a collecting pan
- materials to serve as supports: bricks or wood blocks *(note: these things may get wet)*
- a meter stick
- water source
- rags, paper towels, or sponges for cleaning up

5. Replace the pencil in the hole, refill the jug with water, and set it on the support as you did before. Remove both pencils. Observe how the water flows down the trough.

6. Lower the supports under the trough so that the raised end is only 2 cm above the table surface.

7. Lower the supports for the gallon jug so that the base of the gallon jug is 6 cm above the table (about 4 cm above the raised end of the trough in its new position).

8. Replace the pencils in the holes and refill the jug. Using a single open hole first, and then with both holes open, observe how the water flows down the trough.

9. Does lowering the trough seem to change the flow of the water? Describe what changes (if any) you observed.

____

____

10. Does the water flow differently when you switch from a single hole to two holes? Describe what changes (if any) you observed.

____

____

GUIDE TO ACTIVITY 7

Building a Model of a Stream

◆What is happening?

The purpose of this activity is to allow you to become familiar with the techniques of building and using a simple stream trough. The activity is also designed to demonstrate that direct observations of running water may not provide all the information that we need to answer questions about a stream's flow rate.

Stream troughs can be used to investigate many aspects of the behavior of running water. In subsequent activities, you will learn methods of using your stream trough to examine stream speed, determine how streams carry sediments, and demonstrate how a stream erodes a hillside.

◆Time management

If the jugs have been tested before the activity, one class period should be enough time to complete this activity and clean up. You may need more time if your students have not set up this type of equipment before.

◆Preparation

Plastic gutters are available in most do-it-yourself-oriented home and garden stores. A standard 3.05 m (10 ft) length of gutter can be bought for less than $5.00 in many areas. It will provide enough material for three stream troughs. Plastic gutter is available in several colors. We recommend using white or a light color for these activities, because observing sediments being carried by a stream in Activity 10 is easier if the sediment and the gutter are of contrasting colors.

A mailing tube cut lengthwise and lined with plastic may also be used to perform this activity.

Preparing a stream trough:
Use the handsaw to cut off a 101 cm (1.01 m) length of plastic gutter. Support the gutter so that if the saw slips, no one will be hurt. (Hint: Using a miter box to hold the gutter will allow you to cut it neatly.)

Use sandpaper or a file to smooth the rough edges of the gutter, or cover the the edges of the trough with masking tape or duct tape.

Use the waterproof marker to draw a reference line across the inside of the gutter 1 cm from one end. When the stream trough is being used, this 1 cm reference line will be positioned directly above the edge of the table.

Place your meter stick inside the gutter and find a point 20 cm from the 1 cm reference line that you drew. Use the waterproof marker to draw a line across the inside of the trough at this point.

Measure 50 cm from the second line and draw another line across the inside of the trough.

Check to be sure that the two parallel lines you have drawn are 50 cm apart.

Preparing the jug:
Use a pencil or sharp-pointed scissors to punch two holes near a corner of a 3.8 L (1 gallon) plastic milk jug. The centers of the holes should be located about 1 cm above the jug's base, and about 2 cm apart.

The holes should be small enough so that the jug can be made watertight by sticking pencils into the holes. Test the jug for leaks by placing a pencil in each hole to serve as a stopper, and filling the jug with water.

Remove the pencils from the holes and see if the holes are large enough to allow water to flow freely out of the jug.

Note: If the holes tend to close up when the pencils are removed, the rate at which water flows out of the jug may vary. You can prevent variations in water flow by using sharp scissors to trim the flaps of plastic from around the edges of the holes. The holes remaining should be slightly smaller than the diameter of the pencils that you plan to use.

The four activities in this module using stream troughs require the use of fairly large volumes of water. However, these activities do not require the use of much soil in the troughs, so they are not likely to produce a major mess in the classroom. Even so, you may wish to consider performing the stream trough activities in the schoolyard.

◆Answers

9. Judging the rate of flow of the water using only the naked eye is difficult. We expect the rate of flow to decrease when the slope decreases, but careful observers may report that they are unable to detect any change in the flow when the slope decreases. The next activity will demonstrate a method for determining the speed of the stream.

10. Again, the effects of changing from one to two holes are difficult to observe directly. Some people will report that the stream appears to become wider when the two-hole water source is used.

ACTIVITY 8 WORKSHEET

Calculating the Rate of Flow of a Stream

◆Objective:

In this activity you will learn a technique for determining the speed of water moving along a stream trough.

◆Procedure:

1. Place the trough on a table so that the 1-cm reference line is even with the edge of the table and a short section of the trough extends beyond the edge of the table.

Place supports under the trough so that its raised end is raised 2 cm above the table surface.

Stack supports for the jug near the raised end of the trough. The base of the jug should be about 6 cm above the surface of the table (4 cm above the raised end of the trough).

Use the pencils to plug the two holes in the milk jug. Fill the jug with water and place it on the support with the holes facing the top of the stream trough.

Materials

Each group will need

- the stream trough, collecting pan, 3.8 L jug, pencils, and supports used in Activity 7
- short sections of toothpicks (each about 1 cm long)
- stopwatch capable of measuring 0.01 second intervals
- rags, paper towels, or sponges for cleaning up
- a meter stick

2. Speed is equal to the distance traveled per unit of time. For most scientific purposes, speed is measured in meters per second. Speed can be stated as the equation:

$$\text{Speed} = \frac{\text{distance traveled (in meters)}}{\text{elapsed time (in seconds)}}$$

In order to calculate the speed of water traveling a distance of 0.5 m between the lines drawn inside the stream trough, you need to determine the time that water requires to travel that measured distance. You will determine the time that water requires to move 0.5 m by marking a specific portion of water with a floating toothpick section traveling at the same speed as the water, and observing the toothpick's motion.

You can then calculate the speed of the water as follows:

$$\text{Speed} = \frac{0.5\text{ m}}{\text{seconds it takes for the toothpick to travel between the marks}}$$

Practice the following procedure until the times that you obtain vary no more than ±0.1 second from trial to trial.

Allow the water to begin flowing from one hole of the jug.

Drop a single section of toothpick into the flowing water near the raised end of the trough.

Measure the time that the toothpick takes to move from the upper line drawn inside the trough to the lower line (a distance of 50 cm, or 0.5 m).

3. Now that you have mastered the timing technique, perform three time trials using the single-hole jug with the raised end of the trough elevated 2 cm above the table.

Record the times that you measure for each of the three trials in the data table below. Calculate the average time required, and use that to calculate the average speed of the water in meters per second.

Seconds required for toothpick to move 0.5 meters:

Trough elevation:	**Trial 1**	**Trial 2**	**Trial 3**	**Average time**	**Average speed**
2 cm (1 hole open)	____sec	____sec	____sec	____sec	____m/sec

GUIDE TO ACTIVITY 8

Calculating the Rate of Flow of a Stream

◆What is happening?

Hydrologists make actual stream flow measurements using techniques similar to this toothpick method. These professional scientists sometimes drop dyes or minute amounts of radioactive substances into the stream rather than timing floating objects, but the principle employed is the same.

A toothpick allows you to observe the motion of a specific portion of water moving through a stream, and to measure the time interval required for that marked section of water to move a distance of 0.5 m. Measuring the time required for the water to move this distance allows you to calculate the speed of the stream by using the formula:

$$\text{Speed} = \frac{\text{distance traveled (in meters)}}{\text{elapsed time (in seconds)}}$$

If, for example, the toothpick moves 0.5 m in an average time of 1 second, substituting in the formula for speed gives the following result:

$$\text{Speed} = \frac{0.5 \text{ m}}{1.0 \text{ sec}} = 0.5 \text{ m/sec}$$

At this speed, the toothpick would move 30 m in one minute:

0.5 m/sec x 60 sec = 30 m/min.

It would move 1800 m in one hour:

30 m/min x 60 min = 1800 m/hr.

This is equal to a speed of slightly more than 1 mile per hour, since a mile is equal to 1609 m:

1800 m/hr ÷ 1609 m/mile = 1.1 miles/hr.

◆Time management

Assuming students have done Activity 7 and are familiar with this equipment, one class period should be enough time for taking measurements, making calculations, and cleaning up. You may need to allot more time if students have difficulty with math calculations.

◆Preparation

See "Guide to Activity 7" for instructions on making a stream trough.

Many students have sports watches that can be used to make the time measurements for this activity. Most of these watches have stopwatch functions that indicate intervals of 0.01 sec.; others give 0.1 sec. intervals. Although a stopwatch showing 0.01 sec. intervals is recommended, acceptable results for this activity may be obtained using either type of watch. In most cases, the students' reaction times for starting and stopping the watch are more significant in determining the reliability of data obtained than is the design of the watch.

Accurately timing the toothpick requires some practice. Encourage your students to try this procedure several times before recording their results. They will achieve more consistent results if the toothpick is dropped in exactly the same location each time. Coloring the toothpicks with a permanent marker may make them easier to see as they float down the stream.

You may also want to try using drops of food coloring to measure the speed of the stream. Follow the same procedure as before, but drop the food coloring into the stream instead of dropping the piece of toothpick. The food coloring used to mark a section of water will tend to spread out when it hits the water flowing in the stream trough. The following procedure will give the most reproducible time results:

(1) Start the stopwatch just as the leading edge of the food coloring reaches the upper line.

(2) Stop the stopwatch when the leading edge of the food coloring reaches the lower line.

Since the food coloring tends to spread out, the stock food coloring solution should not be diluted. Dark colors are easiest to see against a white trough.

◆Suggestions for further study

Challenge students to design and build a simple flow meter that will measure speeds at different points along the stream bed. A simple device could be a paddle wheel; advanced students could devise an electronic method for counting revolutions.

If you have a real stream near your school, the activities of this module can be adapted for its use. Safety precautions become a concern when students are near water. Ask parents to help with supervision.

For more sophisticated stream measurement, call your local USGS or other water resource agency for a demonstration of their measuring equipment.

◆Answers

Note: The following data are intended to serve only as a general guideline. Your results may differ from those shown below.

Sample data

Seconds required for toothpick to move 0.5 meters:

Trough elevation:	Trial 1	Trial 2	Trial 3	Average time	Average speed
2 cm (1 hole open)	*0.97* sec	*0.88* sec	*1.01* sec	*0.95* sec	*0.52* m/sec

ACTIVITY 9 WORKSHEET

What Factors Affect the Speed of a Stream?

◆Objective

Using the stream trough as a model stream, you will investigate whether or not varying the slope of the streambed and varying the volume of water passing through the streambed cause changes in a stream's speed.

Materials

Each group will need

- the stream trough, collecting pan, 3.8 L jug, pencils, and supports used in Activity 7
- short sections of toothpicks (each about 1 cm long)
- stopwatch capable of measuring 0.01 sec intervals
- rags, paper towels, or sponges for cleaning up
- a meter stick

◆Procedure

1. Place the trough on a table so that the 1-cm reference line is even with the edge of the table and a short section of trough extends beyond the edge of the table.

Place supports under the trough so that its raised end is 2 cm above the table surface.

Stack supports for the jug near the raised end of the trough. The base of the jug should be about 6 cm above the surface of the table (4 cm above the raised end of the trough).

Use the pencils to plug the two holes in the milk jug. Fill the jug with water and place it on the support with the holes facing the top of the stream trough.

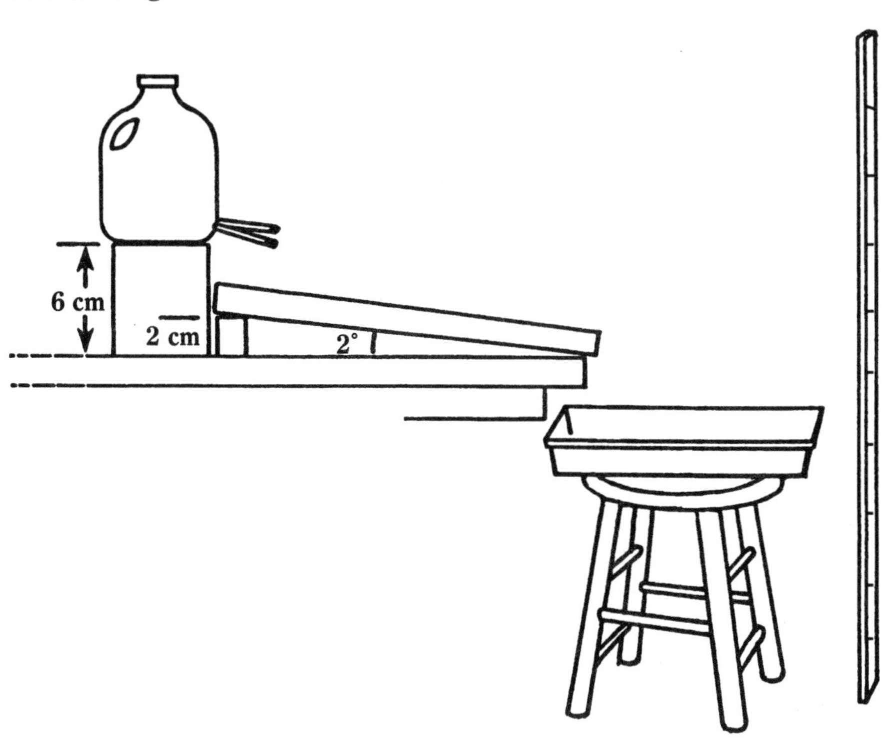

2. Use the following procedure to determine the speed of the stream for each of the conditions listed in the data table.

Allow the water to begin flowing from the jug.

Drop a section of toothpick into the flowing water near the raised end of the trough. For each successive trial, release the toothpick from the same position above the trough.

Seconds required for toothpicks to float 0.5 meters

Trough elevation:	Holes opened:	Trial 1	Trial 2	Trial 3	Average time	Average speed
2 cm	1 hole	____sec	____sec	____sec	____sec	____ m/sec
4 cm	1 hole	____sec	____sec	____sec	____sec	____ m/sec
10 cm	1 hole	____sec	____sec	____sec	____sec	____ m/sec
2 cm	2 holes	____sec	____sec	____sec	____sec	____ m/sec
4 cm	2 holes	____sec	____sec	____sec	____sec	____ m/sec
10 cm	2 holes	____sec	____sec	____sec	____sec	____ m/sec

Measure how long it takes the toothpick to move 50 cm (the distance between the two marks inside the trough).

3. How does the speed of the stream change as the elevation of the trough is increased from 2 cm to 10 cm?

____

____

4. When you open both holes of the jug rather than only one hole, the volume of water flowing into the trough______________ (increases, decreases, stays about the same).

5. How does the speed of the stream change when you open two holes rather than one?

____

____

GUIDE TO ACTIVITY 9

What Factors Affect the Speed of a Stream?

◆What is happening?

The water in a mountain stream flowing down steep slopes travels at a greater average speed than the water flowing in a stream running through a flat valley at the base of a mountain. The general relationship between slope and speed may be stated: As the slope of the streambed increases, the speed of the water flowing in the stream increases. The data that you collected using the 3.8 L (1 gallon) jug water-source should confirm this relationship.

Percent slope is a measure used by engineers, hydrologists, and geologists to measure gradients. The formula for percent slope is

$$\% \text{ slope} = \frac{\text{vertical distance}}{\text{horizontal distance}} \times 100$$

You can calculate the percent slope of the stream trough very easily. The "vertical distance" for the trough is the elevation of the trough above the table (2, 5, or 10 cm). The "horizontal distance" that the water travels while it is above the table is 100 cm for all three elevations. (You do not include the final 1 cm of gutter overhanging the edge of the table as part of the horizontal distance in this calculation.) Therefore, for an elevation of 10 cm,

$$\% \text{ slope} = \frac{10 \text{ cm}}{100\text{cm}} \times 100 = 10\%$$

The percent slope for the 5 cm elevation is 5%, and for the 2 cm elevation, 2%.

If you observe a creek that is carrying more water than usual because of a sudden rainstorm, you may notice that the water also seems to be flowing faster than usual. In fact, the average speed of a stream does increase when the volume of water flowing through the streambed increases. The measurements that you make with the stream trough should confirm this relationship between speed and water volume: The more water that flows down the trough, the faster it will go; therefore, when the elevation of the trough remains constant, the water's average speed will be greater if both holes of the jug are open.

◆Time management

To minimize the number of times that you must reset the slope of the trough, take both the one-hole and two-hole measurements for each trough elevation at the same time.

At least two class periods should be given to this activity—more if students have problems with math calculations.

◆Preparation

See Activities 7 and 8 for information about building and setting up the stream troughs and measuring stream speed.

Calculating average time and average speed:
You may need to explain to your students how to calculate the average time for the different trough elevations. Remind them that they must add the data for Trial 1 + Trial 2 + Trial 3, then divide this sum by 3. The

following example calculates the average time of the Sample Data for a 2 cm elevation with one hole open:

$$.85 \text{ sec.} + .88 \text{ sec.} + .91 \text{ sec} = 2.6 \text{ sec} \div 3 = 0.88 \text{ sec.}$$

This is the average time it takes the toothpick to float 0.5 m.

The distance traveled during each trial by the toothpick is 0.5 m (50 cm). This distance divided by the average time gives the average speed:

$$\frac{0.5 \text{ m}}{0.88 \text{ sec}} = 0.5681... \text{ m/sec, which rounds off to } 0.57 \text{ m/sec.}$$

This is the average speed of the water flowing through the trough.

Accuracy in measuring stream speeds:
Small time intervals are difficult to measure using a hand-held stopwatch. Encourage students to practice timing the toothpick until their results for the three trials fall within about 0.1 sec of each other for each setting of trough elevation and number of open holes tested. You should be aware, however, that this degree of reproducibility may be impossible to achieve for steep settings of the trough when both holes are open. As the speed of the water increases, the time required for the toothpick to travel 0.5 m may become too short for students to measure accurately.

Your results will be more consistent if you drop the toothpick from the same position above the trough each time.

◆Suggestions for further study

Place flat pebbles or screen in the trough to create an irregular stream bed. Water should cover the irregularities when flowing. How does the speed of the water compare with the speed of a smooth stream bed?

◆Answers

3. The speed increases as the elevation increases. The data obtained when one hole is open show that the 10-cm elevation setting produces a stream speed that is almost twice as fast as the speed produced by the 2-cm setting of the trough.

4. The volume of water flowing into the stream trough increases when you open both holes of the jug.

5. The speed of water flowing into the stream trough also increases when you open both holes of the water source. With both holes open, the average speed is about 1.0 m/sec at the 2-cm setting. When only one hole is open and the trough is set at 2-cm elevation, the average speed is about 0.6 m/sec.

Sample data

Seconds required for toothpicks to float 0.5 meters

Trough elevation:	Holes opened:	Trial 1	Trial 2	Trial 3	Average time	Average speed
2 cm	1 hole	*0.85* sec	*0.88* sec	*0.91* sec	*0.88* sec	*0.57* m/sec
4 cm	1 hole	*0.55* sec	*0.62* sec	*0.65* sec	*0.61* sec	*0.82* m/sec
10 cm	1 hole	*0.44* sec	*0.33* sec	*0.49* sec	*0.42* sec	*1.2* m/sec
2 cm	2 holes	*0.47* sec	*0.53* sec	*0.49* sec	*0.50* sec	*1.0* m/sec
4 cm	2 holes	*0.45* sec	*0.42* sec	*0.39* sec	*0.42* sec	*1.2* m/sec
10 cm	2 holes	*0.38* sec	*0.33* sec	*0.35* sec	*0.35* sec	*1.4* m/sec

ACTIVITY 10 WORKSHEET

How Can Streams Move Mountains?

◆Background

In this activity, you will investigate how a stream carries its load of sediments. All streams carry sediments including sand, pebbles, dissolved minerals, and organic materials. Flowing water can quite literally "move a mountain" from an inland location to a river delta or into an ocean basin.

◆Objective

To explain how water contributes to the process of erosion and demonstrate four ways that streams carry sediments

◆Procedure

1. Set up the stream trough with the upper end elevated 5 cm above the table surface. Place the collecting pan at the lower end. Adjust the level of the jug supports so that the base of the jug will rest about 10 cm above the table (5 cm above the end of the trough). Place the pencils in the holes in the jug, fill it with water, and set it on the support.

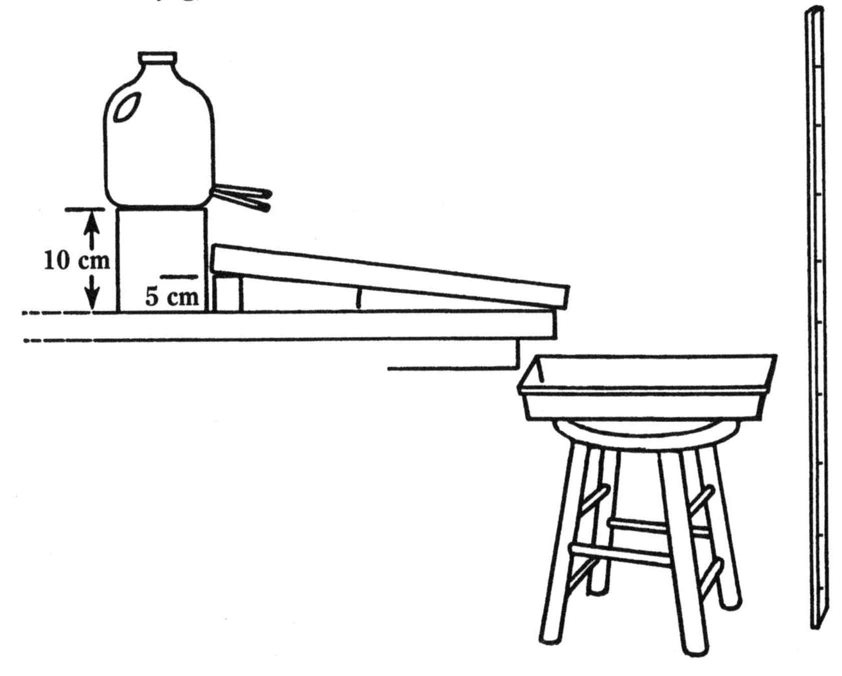

2. Allow the water to begin flowing from one hole of the jug.

Drop a pinch of sand (no more than you can hold between two fingers) into the flowing water near the upper end of the trough and observe what happens. Describe the movement of the sand particles.

____

____

You may see particles of sand bouncing along in the flowing water. This type of movement is called **saltation**. Both wind and water move sand in this way.

Materials

Each group will need

- the stream trough, collecting pan, 3.8 L jug, pencils, and supports used in Activity 7
- sediments: (1) sand (2) round pebbles (3) flat pebbles (4) powdered clay (china clay, kaolin, or pottery clay) (5) ion mixture (a saturated solution of table salt and water)
- an eyedropper
- a mixing jar (paper cup, baby food jar, etc.)
- a stirring rod (a pencil or a plastic straw will work)
- a clean glass microscope slide
- a magnifying glass
- a meter stick
- a water source
- rags, paper towels, or sponges for cleaning up

Note: Your teacher will provide a bucket to collect all wet sediments. Do not dump sediments in a sink—they will clog the drain.

Vocabulary

Colloidal suspension: A method of sediment transport in which water turbulence (movement) supports the weight of the sediment particles, thereby keeping them from settling out or being deposited.

Saltation: The movement of sand or fine sediment by short jumps above the ground or streambed under the influence of a current too weak to keep it permanently suspended.

3. Imagine millions of grains of sand bouncing along in the water of a stream. How might the sand change the streambed?

____

____

4. Would the grains of sand be changed by bouncing along? In what way?

____

____

5. Remove any sediment remaining in the trough after observing the motion of the sand. When necessary, empty your collecting pan and the sediments it contains into the class sediment bucket. Do not pour sediment into the sink—it will clog the drain.

Refill the jug with water whenever necessary.

6. Place four round and four flat pebbles in the upper end of the trough. Allow the water to begin flowing over them.

Describe how the pebbles move down the trough. Are there any differences in the way that round pebbles move compared to flat pebbles? How do you think the shape of the pebbles might change if they were moving down a stream for long distances?

____

____

7. Put a pinch or two of powdered clay in a mixing cup of water and stir vigorously until the mixture appears cloudy. This clay-and-water mixture is called a **colloidal suspension.**

Start the water flowing down the trough from one hole and pour the suspension of clay and water into the stream. How is the colloidal suspension transported by the stream?

____

____

8. Use the dropper to add a small amount of the salt solution to the upper end of the flowing stream of water.

Can you observe the salt solution being carried by the stream? If so, describe how it is carried by the stream.

____

____

9. One way to tell whether or not a stream is carrying dissolved materials is to obtain a water sample and allow the water to evaporate. If salts are present in the water, they will crystalize as the water evaporates. Test for the presence of salt in the ion mixture as follows:

Place two drops of the ion mixture on a clean microscope slide. Set the slide in a warm place and allow the water to evaporate. Use a hand magnifying lens (or microscope, if one is available) to observe what remains after the water evaporates. Sketch or describe your observations:

10. List four methods by which streams move sediments.

1.____

2.____

3.____

4.____

11. Complete the following statements:

A faster-moving stream will be able to carry ________________ (*more, less, the same amount of*) sediments as/than a slower stream.

A faster-moving stream will be able to carry ________________ (*larger, smaller, the same size*) sediments as/than a slower stream.

GUIDE TO ACTIVITY 10

How Can Streams Move Mountains?

◆What is happening?

Sediments are the fragments of rocks, minerals, and organic material produced by the forces of weathering and erosion. Sediments vary in size from huge boulders moved by flooding streams to the atom-sized minerals dissolved as ions or salts in stream water.

Fast, rain-swollen streams carry heavy sediment loads; wherever they slow down, part of that load is deposited in the streambed. The largest particles fall to the bottom first; smaller particles suspended in the water eventually settle to the bottoms of lakes or ocean basins. All streams carry sediments in an endless trip from dry land to the ocean basins, where the lower layers of sediments are slowly transformed back into rock by the pressure of overlying sediments.

Streams are important agents of erosion and are constantly lowering and leveling the land above sea level. Waterborne sediments are the tools of the streams, carving out valleys and canyons as they move along. Anyone observing the smooth, rounded rocks in a swift-running mountain stream may also infer (correctly) that the tumbling and scraping tends to smooth and round the sediments themselves as they move downstream.

While performing this activity, you observed four ways that sediments can be transported:

1. in solution (the salt dissolved in the water)
2. as a colloidal suspension (the clay particles that are too small to see individually, but large enough to give the water a milky appearance)
3. by saltation (the bouncing motion of the sand particles)
4. by rolling and sliding (in real streams, large rocks move this way)

◆Time management

Assuming students have done previous activities, this activity should take about one-and-a-half class periods to complete (including cleanup). Follow-up discussion should take another one-half to one period of class.

◆Preparation

Ideally, the sand, round pebbles, and flat pebbles should all be quartz or materials of similar density. The point of this activity is to vary particle size, *not* composition.

Mix a saturated salt solution as follows: Add salt to 100 ml of water and stir. Continue adding salt and stirring until no more salt dissolves. Allow the undissolved salt to settle on the bottom of the container. Pour off the clear salt solution above the salt granules without disturbing the undissolved salt.

If you do not have ready access to dry clay from which to make the colloidal suspension called for in step 7, you can make an acceptable substitute as follows:

Take a fist-sized clump of soil that contains some clay, break it up into small pieces, and place it in about 500 ml of water. Stir the soil and water vigorously. The smallest particles of the soil (the clay components) will become suspended in the water, making it appear cloudy. The particles in this type of mixture are less than 1/256 mm in diameter. They settle out of the water extremely slowly unless chemicals (such as alum) are added to clump them together.

Pour off the top layer of cloudy water from the container, and use it in step 6. When decanting the cloudy water, avoid disturbing the larger particles of soil that sink to the bottom. Skim off any organic material that floats to the top.

You may want to discuss the definitions of clay, sand, and pebbles with your classes before performing this activity. The differences among these are explained in the introduction to Module 2.

◆Suggestions for further study

Have students check their answers to question 10 by designing a brief experiment to investigate how the speed of the stream affects the sediment that it can transport.

One such experiment might be set up as follows:

1. Measure out two equal quantities of sediments (consisting of sand, soil, and small pebbles).
2. Set the stream trough at a "low" setting (2 cm above the table) so that it will have a slow rate of flow.
3. Place one of the sediment samples at the upper end of the stream trough and time how long it takes the water coming from the one-hole jug to carry the sediments to the bottom of the trough.
4. Reset the stream trough at a "high" setting (10 cm above the table), increasing the speed of the water flowing through it.
5. Place the second sediment sample at the upper end of the stream trough and time how long it takes the water coming from the one-hole jug to carry the sediments to the bottom of the trough.
6. Compare the times required to move the sediments for the "slow" and the "fast" streams.
7. Students may also want to determine the speeds needed to carry specific-size sediments or sediments of different densities.

Collect water samples from a local stream at regular intervals and compare the amount of sediment that settles from each. Discuss why the amount of sediment varies.

◆Answers

2. The grains of sand bounce down the trough when the water flows over them.

3. Millions of sand particles bouncing along a streambed will tend to smooth out the streambed and any large rocks lying in it.

4. The sand grains themselves will tend to become smoother and rounder as a result of rubbing against the streambed and each other.

6. Round pebbles tend to roll down the trough. Flat pebbles slide along. As rocks slide and tumble over one another and through abrasive sediments, they tend to develop smooth surfaces and rounded edges. The smooth, rounded gravel found in fast-moving mountain streams is produced in this manner.

7. Particles in colloidal suspensions are carried along suspended in the moving water of a stream. They move at the same speed as the water.

8. All running water contains salts (also called dissolved minerals or ions). These substances are usually present in concentrations that are too low to detect without evaporating the water or performing a chemical analysis of it. Visually detecting the presence of ions is usually impossible; individual ions are much too small to see and most ions being carried by streams form colorless solutions at normally encountered concentrations. Copper sulfate and some other crystals form brightly colored solutions when they dissolve, but even these brighly colored ionic solutions rapidly disperse and "disappear" when they are diluted by flowing water.

9. The sodium chloride (NaCl, or table salt) used to prepare the ion solution will form cubic crystals.

10. Sediments can be transported:

• in solution (the salt dissolved in the water)

• as a colloidal suspension (such as clay particles that are too small to see individually, but give the water an overall milky appearance)

• by saltation (the bouncing motion of the sand particles)

• by rolling and sliding (in real streams, larger sediments may also move this way).

11. A faster-moving stream will be able to carry *more* sediments than a slower stream.

A faster-moving stream will be able to carry *larger* sediments than a slower stream.

ACTIVITY 11 WORKSHEET

How Ice Shapes the Land—A Glacier on the Rocks

◆Background

In the United States, very few people are aware of the great changes in the landscape caused by ice. Yet many of our most prominent geological features, including the Great Lakes, formed as a result of the "rivers of ice" called glaciers that once covered the northern half of what is now the United States.

◆Objective

To explain how water contributes to the process of erosion and show how the abrasive action of glaciers can change the surfaces that they flow across

◆Procedure

1. Drop the irregularly shaped pebbles and the spoonful of sand into one of the margarine tubs.

2. Fill both of the tubs about three-quarters full of water.

3. Place both tubs in a freezer until the water in them is frozen solid. (This will take several hours.)

4. The next day:

Remove the plain ice from the margarine tub. Wet the bottom surface of the ice or allow it to melt a little before placing it on the brick.

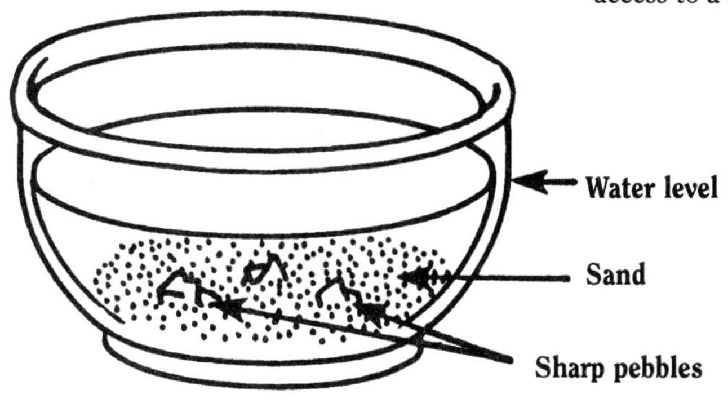

Place the plain ice on a smooth surface of the brick, press down hard, and slide it back and forth across the brick.

Describe how the ice moves across the brick. Does it scratch the brick's surface or remove any material from the surface of the brick?

____

____

5. Now remove the rocky ice from its tub, wet its lower surface or allow it to melt slightly, and place it on the brick with the sand and pebbles touching the brick. Press down hard, and slide it across the brick in the same manner that you did with the plain ice.

Describe how the rocky ice moves across the brick. Does it scratch the brick's surface or remove any material from the surface of the brick?

____

____

Materials

Each group will need

- three or four small (pea-sized) irregularly shaped pebbles
- a spoonful of sand
- 2 plastic soft margarine tubs or plastic bowls of similar size
- a common, unglazed brick, brushed and rinsed to remove any loose particles
- water
- access to a freezer

6. What do you think might happen to the pebbles if you continued to move them back and forth across the brick?

7. Imagine that the rocky ice is part of a glacier 100 m thick (as thick as a 30-story building is tall). In addition to sand and pebbles, this glacier has huge amounts of dirt, large rocks, and boulders larger than automobiles locked in its ice and scraping along the ground as the glacier advances.

Describe what you think might happen as this rock-carrying glacier moves across the Earth. What type of marks would it leave on the Earth's surface?

8. Place the rocky ice on a surface that will not be harmed by water, and let it melt. What happens to the sediments (sand and pebbles) it is carrying?

GUIDE TO ACTIVITY 11

How Ice Shapes the Land— A Glacier on the Rocks

◆What is happening?

This activity provides a brief demonstration of the abrasive ability of frozen water (glaciers) carrying rocks and sand. It can be used in the classroom to introduce the concept that glaciers cause great changes on Earth's surface.

During long periods of cold climate, glaciers expand as they accumulate ice and snow, and advance. During warm climate periods or years during which little snow falls, the glacial ice melts and glaciers retreat. As glaciers advance, rocks and smaller sediments frozen in them are dragged along. Sediments near the bottom and sides of glaciers scratch and gouge materials that they slide across. This abrasion smooths surface rocks, rounds hills, and scours out valleys. On some exposed rock surfaces you can find evidence of this scouring in the form of parallel lines called **glacial striations** carved into the rock.

Glaciers, like rivers, tend to follow the course of least resistance. They generally move downhill and flow around obstacles that are too large to push aside. Gravity pulls the ice of alpine glaciers into the valley systems of rivers that existed before the glacier formed. The abrasive action of glaciers on the sides of valleys changes the typical "V" shaped river valleys into "U" shaped glacial valleys.

When glaciers melt, they dump mixtures of sediments on the land. Boulders called **erratics** are among the most striking calling cards left by retreating glaciers. Erratics may be as large as houses. These huge boulders stand out sharply from the flat areas where they were deposited.

Glacial **till** may provide geological evidence for a glacier's migration. The materials found in till can often be traced to their sources hundreds of miles north of the location where they are found. Much of the fertile soil in the northern plains of the United States was a glacier-borne "gift" from Canada, transported to the U.S. by glaciers.

Vocabulary

- **Erratic:** Huge boulders, often carried for long distances by advancing ice sheets, that are dropped by melting glaciers and remain on the glaciated landscapes.
- **Glacial striations:** Lines carved into rock by overriding ice, showing the direction of glacial movement.
- **Till:** A deposit of sediment formed at the base of a glacier, consisting of an unlayered mixture of clay, silt, sand, and gravel ranging widely in size and shape.

◆Time management

Rather than making the plain ice and rocky ice blocks in class, you can make them yourself a day or two before the workshop. The procedure for preparing the ice blocks is given in steps 1–3 of this activity.

Procedure steps 1–3 should take about 15 minutes if the materials are readily available to the students. The next part of the procedure should take about half a class period to perform, except for step 8 which may take a few hours and will most likely have to be observed for a few minutes the next day.

◆Preparation

Students can perform this as a hands-on activity. However, if your freezer space is limited, you may find it more practical to do this activity as a demonstration for your classes. If so, have student assistants do steps 1 and 2 on one day, and steps 4, 5, and 8 on the following day. Be sure to pass the rocky ice around the class so that all students get to feel how rough it is.

If there is any loose clay or dirt on the surface of the brick, the plain ice may look dirty after rubbing against the brick. This may be interpreted as evidence for the ice removing part of the brick. Rinsing the brick with water before the demonstration will prevent this confusing result.

◆Suggestions for further study

Test the ability of your rocky ice sample to scratch the surface of samples of soft rocks such as slate and sandstone. You can also use a piece of smooth wood to demonstrate the abrasive ability of the ice containing rocks and sand.

◆Answers

4. The plain ice slides smoothly across the face of the brick.

5. The rocky ice scrapes off part of the surface of the brick. The ice will have brick-colored clay particles on its lower surface trapped among the sand and pebbles.

6. Scraping the pebbles embedded in the ice across hard surfaces will flatten the exposed lower side of the pebble. Rocks having one side that is flattened in this manner are called faceted rocks. They are often found in glacial deposits.

7. Glaciers leave gouges and scrapes in the land. In valleys that once contained glaciers, exposed rock surfaces will exhibit marks that run parallel to the direction of the glacier's movement.

8. When the ice melts, it drops its load of sand and pebbles. Retreating glaciers do the same thing as they melt, depositing huge rocks and boulders. Streams from the melting ice may also move rocks and pebbles around and redistribute the sand.

MODULE 3

Raindrops Keep Falling on My Head

◆Introduction

- In order for 2.5 cm of rain to fall on a football field, about 117,000 kilograms of water must fall from a cloud.
- According to studies done by the U.S. Soil Conservation Service, raindrops splash between 900 and 90,000 kg of soil into the air per acre of farmland during each rainstorm.
- Acid rain has increased the average acidity of lakes in the Adirondack Mountain region of New York by a factor of 100 since the 1930s. Many of these lakes are now almost as acidic as orange juice and can no longer support typical lake life.

Farmers worry about having too much or too little rain. Most of us know that the farm economy suffers during periods of drought, but not everyone is aware that much of the farmland in this country is subject to critical damage from erosion caused by the effects of rainfall.

Most of us look forward to gentle spring rains; but in some parts of the United States, spring rains contain so much acid that they can be deadly to the fish and plants that they fall upon.

This module examines topics related to rain: how clouds form, how rain affects the surface of the ground, how scientists measure the amount of acid in liquids, and what problems are produced by acid rain.

SUMMARY OF MODULE 3

Raindrops Keep Falling on My Head

◆Instructional objectives

After completing the activities and readings for Module 3, you should be able to

- describe how clouds form [Activities 12 & 13]
- make a small cloud in the laboratory [Activity 13]
- demonstrate how mulching and contour farming practices can reduce erosion [Activities 14 & 15]
- measure the erosion occurring on a model of a cultivated field. [Activity 15]
- demonstrate how certain types of soil can neutralize acid rain [Activity 16]

◆Preparation

Study the following readings for Module 3:

Reading 10: The Invisible Gas Becomes Visible: Clouds

Reading 11: How Acidic Is the Rain?

Reading 12: Soil Erosion: The Work of Uncontrolled Water

◆Activities

This module includes the following activities:

Activity 12: Film: "What Makes Clouds?"

Activity 13: Put a Cloud in a Bottle

Activity 14: How Raindrops Erode the Soil

Activity 15: How Can Farmers Reduce Erosion Caused by Rain?

Activity 16: How the Soil Affects Acid Rain

ACTIVITY 12: FILM

Encyclopaedia Britannica Film: "What Makes Clouds?"

◆Background

This film uses both laboratory demonstrations and photographs of natural clouds to explain the conditions required for cloud formation. It relates cloud formation to the processes of evaporation, condensation, and transpiration, and answers questions such as:

1. What are clouds composed of?
2. What conditions are needed for clouds to form?
3. What determines how much water vapor air can hold?
4. Why does rising air expand?

◆Objective

To understand how clouds form

◆Time management

The running time of this film is 19 minutes. One class period should be allotted for an introductory lecture, viewing, and discussion of the film.

◆Summary of film content

For invisible water vapor to condense into visible water droplets, two conditions must exist: saturated air must be cooled, and there must be surfaces upon which the droplets can form.

As portions of air near the Earth are heated by the sun, they start to rise. When warm air rises, it expands and cools. If water vapor is present in the air, condensation will occur and clouds will form. Condensation surfaces involved in producing cloud droplets are tiny, microscopic particles of dust, smoke, or salt that are light enough to remain suspended indefinitely in air.

The film includes abundant visual evidence for this explanation of cloud formation. In addition to a laboratory experiment illustrating the saturation of air and condensation of moisture, there are extended time-lapse sequences showing clouds forming in the sky, and several examples of transpiration, evaporation, and condensation. A dramatic answer to the question of whether air expands as it rises above the Earth is provided in scenes showing the ascent of a plane carrying an inflatable bag attached to a plexiglass box. As the plane climbs to 10,000 feet above sea level, air in the box expands and fills the bag. Problems suggested at the close of the film indicate several important areas for individual follow-up study: the cause of fog, the nature of high altitude clouds, and the formation of raindrops.

Source: Encyclopaedia Britannica Educational Corporation study guide for film.

ACTIVITY 13 WORKSHEET

Put A Cloud in a Bottle

Materials

Each group will need

- a clear plastic soft drink bottle and its cap (use a 1 or 2 L bottle that is clear, not tinted green)
- a small beaker of water (50 ml of water is ample)
- matches
- light source—a flashlight or a small lamp

Vocabulary

- **Cloud:** A mass of suspended water droplets and/or ice crystals in the atmosphere.

◆Background

Two conditions must be present to form a **cloud:**

1. The air must be saturated with water vapor.
2. The air must contain particles for the water vapor to condense on.

◆Objective

To demonstrate how clouds form by making a small cloud in the laboratory

◆Procedure

1. Remove all labels from the bottle.

2. Pour just enough water into the bottle to cover the bottom. Screw on the cap, and shake the bottle vigorously 20 times.

3. Open the bottle and quickly pour out most of the water. Leave about a capful of water inside the bottle.

4. Screw the cap securely onto the bottle. Hold it beside the light source so that the light shines through the bottle at a right angle to your line of sight.

5. Squeeze the bottle hard, using both hands, then relax your grip. (Note: If this forces any air out of the bottle, tighten the cap again and repeat this step.) Do you see any evidence of cloud formation inside the bottle? Describe any changes that you see.

____

____

____

6. Repeat steps 2 and 3.

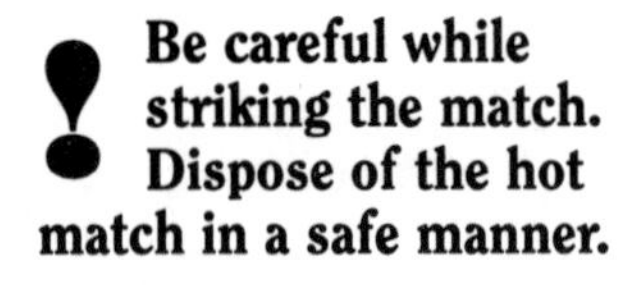

Be careful while striking the match. Dispose of the hot match in a safe manner.

7. Strike a match, let it burn briefly, then blow it out. Quickly move it near the mouth of the bottle so that the smoke from the match goes into the bottle. You can get the smoke into the bottle by:

- gently blowing the smoke toward the bottle, or
- sucking the smoke into the bottle. Squeeze the sides of the bottle, bring it near the smoke, and release the sides. Air rushing back into the bottle will pull the smoke in.

As soon as you can see a small amount of smoke inside the bottle, screw the cap on securely.

8. Place the bottle near the light as you did in step 4. Squeeze the bottle hard, then relax your grip. Do you see any evidence of cloud formation inside the bottle? Describe any changes that you see.

____

____

9. Squeeze and release the bottle again. Describe any changes that you see.

____

____

10. Why are the results different when there is smoke inside the bottle?

____

____

GUIDE TO ACTIVITY 13

Put a Cloud in a Bottle

Vocabulary

- **Condensation surfaces:** Small particles of matter, such as dust and salt suspended in the atmosphere, which aid the condensation of water vapor in forming clouds; also called condensation nuclei.

◆What is happening?

Shaking water inside the bottle helps to saturate the air inside with water vapor. You cannot observe this directly, however, because water vapor, the "raw material" of clouds, is invisible.

Squeezing the bottle increases the pressure and heats the gases inside. Warm air can hold more water vapor than cold air, so additional water vapor enters the air inside the bottle.

Relaxing your grip decreases the pressure and cools the gases. If there are particles present in the air inside the bottle, the water will condense on these particles as it expands, forming the visible droplets that we call a cloud. Smoke particles are very efficient condensation surfaces—able to facilitate cloud formation at room temperature.

◆Time management

One class period should be enough time to perform this activity and discuss the few questions.

◆Preparation

You may wish to ask each group performing this experiment to provide its own flashlight and a 1 or 2 L bottle with a tightly fitting cap.

The flashlight is not essential for this activity, but transilluminating the bottle makes the cloud much easier to see. If the room lights are dimmed or turned off, the cloud can be seen more easily. A clip-on lamp will also work.

◆Suggestions for further study

To study airborne particles that can serve as condensation nuclei, cut 1 cm x 1 cm holes in several 3 x 5 cards. Cover the holes with tape. Place cards outdoors with the sticky side of the tape facing out. The cards should be in a variety of locations, at different heights, and facing different directions. Record all data on the card. Collect the cards after several days and seal in plastic bags. Examine under a microscope.

◆Answers

5. No visible change will occur when the bottle is squeezed. Be sure the bottle is tightly sealed when the cap is screwed on.

8. After smoke particles are introduced into the bottle, the bottle fills with a cloud when it is squeezed hard and then rapidly released.

9. Using this apparatus, clouds can be formed and made to disappear repeatedly. When the bottle is squeezed, the cloud inside disappears. Releasing the pressure on the bottle allows the air inside to cool slightly, and the cloud re-forms.

10. In order for a cloud to form, molecules of water vapor must come together and condense around a particle of some kind that is suspended in the air. Such particles are called **condensation surfaces.** In this activity, the smoke particles that you put into the bottle serve as condensation surfaces. In nature, smoke, dust, pollen, meteoric dust, and salt particles from ocean spray often serve as condensation surfaces.

ACTIVITY 14 WORKSHEET

How Raindrops Erode the Soil

◆Background

Raindrops strike the ground with enough force to dislodge soil, especially if the soil has been stripped of its natural cover of vegetation. Water running across the land can carry away soil particles loosened by rain. In this activity, you will use a model of a hillside to investigate how raindrops might affect the ground's surface features.

◆Objective

To demonstrate how mulching and contour farming practices can reduce erosion

◆Procedure

1. Spread out newspapers or plastic sheets to protect any surfaces that might be damaged by water or soil.

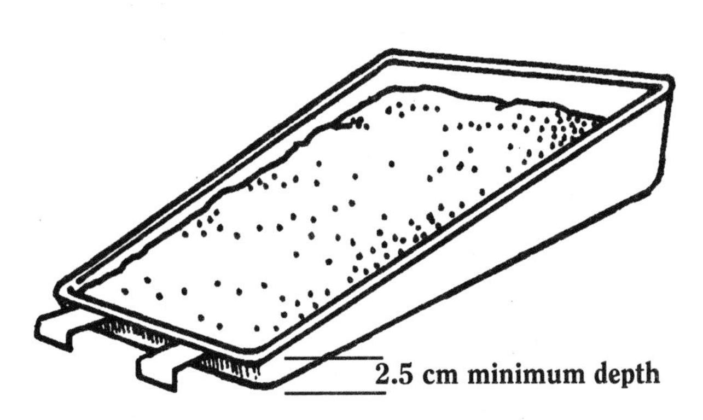

2. Set up a model of a hillside as follows: Fill the paint roller tray with soil. At the shallow end of the pan, the top of the soil should be just below the rim and have a minimum depth of about 2.5 cm. (The bottoms of these trays are uneven, so the maximum depth will vary.) The surface of the soil should be smooth and level, and parallel to the top edge of the tray.

3. Make two stacks of bricks or wooden blocks to elevate the model hillside above the plastic basin. The lower end of the hillside should hang over the plastic basin. The rim of the raised end of the hillside should be 6 to 7 cm above the rim of the lower end.

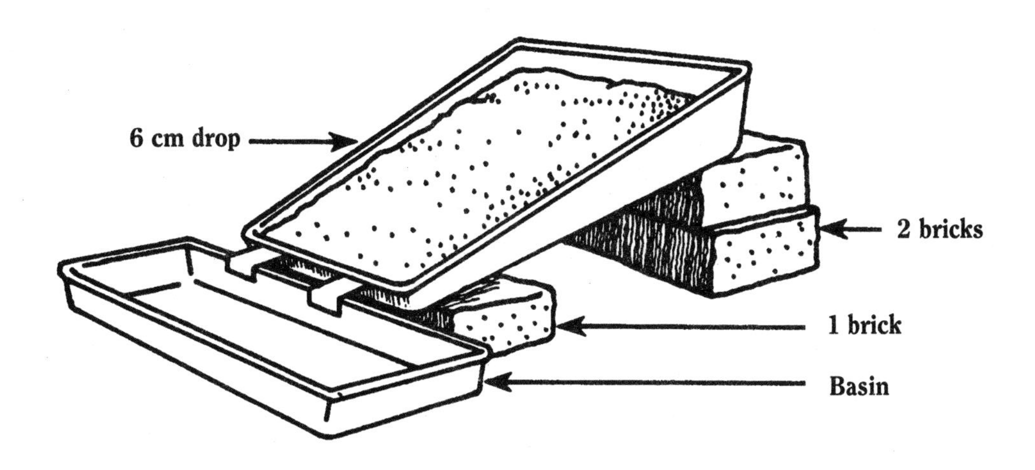

Materials

Each group will need

- a paint roller tray
- a collecting basin that is wider than the paint tray
- a sheet of notebook paper
- scissors
- a watering can (a beach toy is fine)
- topsoil—enough to partially fill the roller tray
- mulch—grass clippings, straw, leaves, or shredded paper
- bricks or wooden blocks for supporting the roller tray
- a water source
- a meter stick
- newspapers or plastic sheeting
- rags, sponges, or towels for cleaning up spills
- two 500 ml beakers or paper cups
- a clock or wristwatch

Vocabulary

- **Erosion:** The processes of picking up sediments, moving sediments, shaping sediments, and depositing sediments by various agents. Erosion plays a role in creating Earth's surface features—the landscape. Erosional agents include streams, glaciers, wind, and gravity.

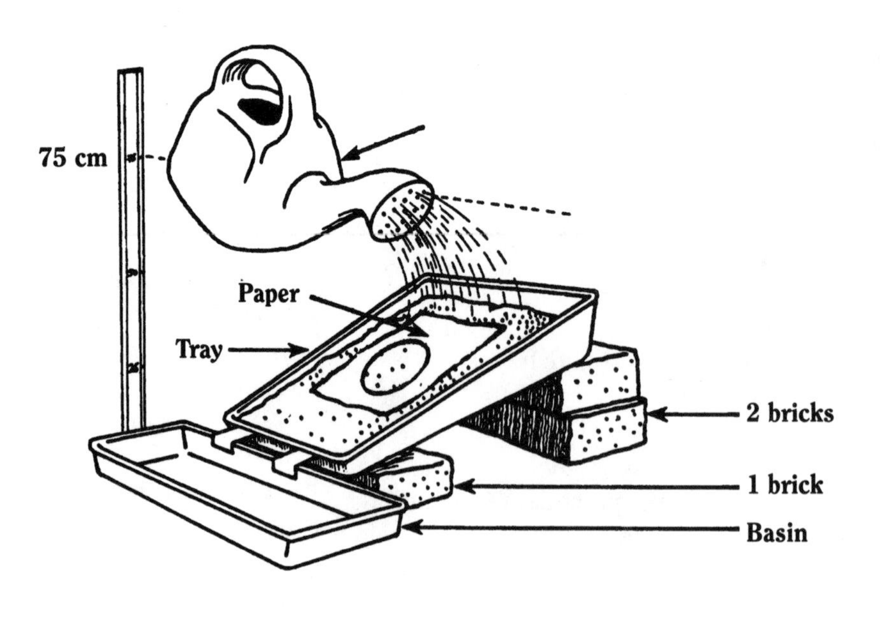

4. Cut a circular hole in the sheet of notebook paper. The hole should be about the same diameter as the sprinkling head on the watering can.

5. Place the sheet of paper on top of the soil of the model hillside. Use the watering can to pour water from a height of 75 cm through the hole onto the soil for several seconds.

6. Is there evidence that the raindrops from the watering can are moving soil particles? Describe your observations of the soil and the paper.

____

____

____

____

____

____

7. Remove the paper from the model hillside and smooth out the soil.

8. Pick up the lower end of the model hillside, and hold it so the soil surface is level (rather than sloping toward the plastic basin).

Pour water evenly over the entire surface of the bare soil until it appears to be thoroughly soaked. Let it absorb water for three minutes. Add water if the surface appears to dry out.

Put the model hillside back in the slanting position and allow any excess water remaining in it to drain into the plastic basin.

9. Empty the water in the basin into the class collection basin. Do not pour water containing soil directly into the sink—the soil will clog the pipes and prevent the sink from draining properly. Scrape out any sediments remaining in the bottom of the basin and discard them as directed.

10. Refill the watering can. Simulate a rainstorm by rapidly sprinkling water on the upper end of the model hillside for 10 to 20 seconds from a height of 75 cm. Describe how the water moves across the soil.

____

____

____

11. **Erosion** is occurring if soil is being carried away by the action of the falling water. Describe how the surface of the soil was changed by the rainstorm.

____

____

____

12. Carefully pour off the water in the basin into the class collection basin; be sure not to pour off the soil.

13. Place the soil that washed from the model hillside into the basin in a beaker, and estimate (in milliliters) how much soil the model hillside lost during the rainstorm.

Volume of soil lost = _______ ml.

14. Rinse out any sediments remaining in the bottom of the basin.

Replace the soil that washed off the model hillside and smooth the surface of the model hillside so that it looks the same as it did before the rainstorm (you may have to add a little extra soil).

Now repeat steps 8 and 9.

15. Cover the surface of the soil with a layer of mulch. The mulch should cover the soil much the way that leaf litter covers the soil in a forest.

16. Refill the watering can and rapidly sprinkle water on the raised end of the mulch-covered model hillside for 15 seconds from a height of 75 cm. Describe how the water moves across the mulch-covered soil.

____

____

____

17. Repeat step 12.

18. Collect the soil that washed off the mulch-covered model hillside into the basin below. Place this soil in a beaker and estimate (in milliliters) how much soil the model hillside lost during the rainstorm.

Volume of soil lost = _______ ml.

19. Is erosion reduced by covering the soil with mulch? Compare the volume of soil washed off the mulch-covered model hillside with the amount washed off when the soil was bare.

____

____

____

GUIDE TO ACTIVITY 14

How Raindrops Erode the Soil

◆What is happening?

Using a paint tray filled with soil as a model hillside is an excellent way to illustrate how water running across sloping land causes erosion. After working with this apparatus, most people find it much easier to understand how gullies form and how other types of erosion occur.

Pouring the "rain" through the hole in the paper demonstrates the process of rain splash erosion. Even these simulated raindrops land with sufficient force to dislodge soil particles and move them onto the paper. Real raindrops fall from much greater heights and hit the ground with much more force, moving even larger soil particles.

This activity also shows how rain splash can lead to other types of erosion such as (gullying) and sheet wash, in which a thin layer of water flows over an entire surface. The soil dislodged by raindrops can be carried rapidly to streams if the soil surface is bare and runoff water is allowed to travel unimpeded across sloping ground. Covering the soil with a layer of mulch can greatly reduce the amount of erosion occurring by cushioning the impact of the raindrops and slowing down the runoff of surface water.

The United States has one of the most extensive supplies of potentially productive soil in the world. Our farmers have exploited this resource, making this country one of the world's largest food producers and exporters. However, the amount of soil being lost each year because of erosion greatly exceeds the amount of new soil being formed. If this trend is not reversed, our economic well-being and international political power may decline in future years.

Much of the erosion in the United States occurs as a result of human activity. Plowing fields for farms, cutting trees for lumber, clearing house lots, and even recreational land use such as hiking along mountain trails may leave the soil exposed and vulnerable to erosion. All of these activities are valuable; but if they are done carelessly, or in erosion-prone areas, they can cause soil loss.

◆Time management

This activity will take between one-and-a-half and two periods to complete and clean up.

If you have several consecutive classes that will be performing the activities, you may wish to do the erosion activities with every other class on two different days. Skipping a period between classes performing laboratory activities will allow you some time for cleaning up and reorganizing the materials. The classes that do not perform the activity on the first day can be given an alternative assignment to do.

◆Preparation

To cut down the volumes of soil that the teacher must gather, transport, and dispose of, you may wish to do this activity as a demonstration. You can maintain the hands-on aspect by having groups of students perform the demonstration for their classmates. The students will be able to see the demonstrations better if you construct a larger model hillside, or purchase one from a scientific supplier. (These are sometimes listed in catalogs as "stream tables.")

You may want to measure the volume of soil it takes to fill a tray in order to compute the percentage of soil eroded into the basin.

However, we strongly recommend having the students perform this activity and "How Can Farmers Reduce the Erosion Caused by Rain" as hands-on exercises if at all possible. The following strategies will make doing these activities easier for you and the students:

Before starting this activity, a collection basin for the water that contains soil particles should be set up.

Do the activities outside. This will reduce cleanup to a minimum. If possible find an area that is near a water source (such as a garden hose).

Any good, loose garden soil (preferably one that does not have a high clay content) may be used for this activity. You may be able to dig up dirt from the schoolyard for use in the trays and then dispose of the wet dirt by simply filling in the holes. It is probably a good idea to screen the soil and remove any large stones or clumps before starting the activity.

Alternatively, you may use commercially available potting soil. If you use potting soil, test a small batch of it before starting the activity to be sure that it absorbs water readily and is neither too loose nor too clumpy to use.

A beaker or cup makes a good template for the circle on the paper. You could also trace the sprinkling head of the watering can.

Both this activity and Activity 15 use the same apparatus. If you do not have enough watering cans to use for this activity, you can make a sprinkler by using a small nail to punch holes in the bottom of a coffee can or other large can. This type of sprinkler works very well. The plastic top from the coffee can be used to stop the flow of water. You can also use a plastic soda or milk container with a cap. Punch holes near the top on one side and invert to use after the cap is replaced. You may wish to ask each group to provide its own watering device.

Old bath towels are excellent for cleaning up after this type of activity.

◆Suggestions for further study

Set up a demonstration garden. Mulch one side of the garden, and leave the soil bare on the other side. Observe the growth rates of plants of the same species in the two different areas of the garden. Collect samples of runoff water during a rainstorm, and compare the amount of soil carried away by water on the mulched versus the unmulched side of the garden.

After your students have learned about the causes and solutions for erosion, take them on a field trip around the schoolyard and let them examine actual examples of erosion. Most schools have several areas that are eroding—even if the erosion is occurring only in the areas near downspouts from the roof. You may even wish to have your students do an erosion control project for the school.

Invite guest speakers from the local soil and water management authority to talk about sedimentation problems in reservoirs and streams. Erosion and loss of prime soil is only part of the problem caused by uncontrolled runoff. The damage that displaced sediment causes to waterways, reservoirs, and stream channels is expensive to remedy. Accumulation of sediments can increase the frequency of flooding events because the sediment takes up storage space in reservoirs and stream channels.

Have the students vary how high the watering can is held. How does this affect the amount of erosion and the pattern of sediment on the paper?

◆Answers

6. A depression forms where the water strikes the soil. Particles of soil are splashed onto the surface of the paper.

10. A sheet of water (called sheet wash) moves across the surface of the soil, carrying a large amount of soil to the lower end of the tray or into the basin. Some channelized flow (gully erosion) may also be visible.

11. The water droplets cause a depression to form in the upper end of the tray. Small gullies may form.

12. Answers will vary depending on the type of soil used, the force of the water, and the size of the tray. In several trials, we estimated that 25% of the soil in the tray was dislodged by the water.

15. The mulch breaks up the water flowing across the surface. The water does not flow down the pan in a continuous sheet. The mulch also slows down the rainfall and breaks the water into smaller droplets.

16. Answers will vary depending on the type and thickness of the layer of mulch, the type of soil used, the size of the holes in the sprinkler head, and the size of the tray. Less than 5% of the soil in the tray was dislodged from the mulched hillside in several of our trials.

17. Much less erosion takes place when the soil is covered by mulch than when the bare soil is sprinkled. The soil beneath the mulch stays firm and relatively undisturbed.

ACTIVITY 15 WORKSHEET

How Can Farmers Reduce Erosion Caused by Rain?

◆Background

Water running across land being used for growing crops can carry away large amounts of topsoil. In this activity, you will investigate how different plowing techniques affect the rate of erosion.

◆Objective

To demonstrate how mulching and contour farming practices can reduce erosion, and to measure the amount of erosion occurring on a model of a cultivated field

Before starting this activity, a collection bucket for the water that contains soil particles should be set up. All groups must dispose of their sediment-containing water in this bucket.

◆Procedure

1. Spread out newspapers or plastic sheets to protect any surfaces that might be damaged by water or soil.

2. Set up two model hillsides as follows: Fill the paint roller trays with soil. At the shallow end of each pan, the top of the soil should be just below the rim and have a minimum depth of about 2.5 cm. (The bottoms of these trays are uneven, so the maximum depth will vary.) The surface of the soil should be smooth and level, and parallel to the top edge of the tray.

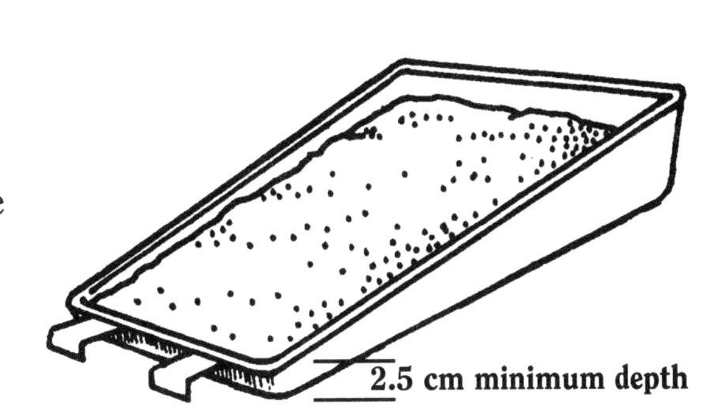

3. Make two stacks of bricks or wooden blocks to elevate each model hillside above the plastic basins. The lower ends of the hillsides should hang over the plastic basins. The rims of the raised ends of the hillsides should be 6 to 7 cm above the rims of the lower end.

4. The model hillsides can be used as models of fields plowed in different ways as follows:

Using a pencil as your plow, make a series of parallel ditches about 2 cm deep in the soil of both hillsides. On one hillside, make the ditches go directly down the slope; on the other, make the ditches go across the slope. See the diagram on the next page.

Materials

Each group will need

- 2 paint roller trays
- 2 collecting basins that are wider than the paint trays
- 1 watering can
- topsoil—enough to partially fill the roller trays
- bricks or wooden blocks for supporting the roller trays
- a water source
- a meter stick
- newspapers or plastic sheeting
- rags, sponges, or towels for cleaning up spills

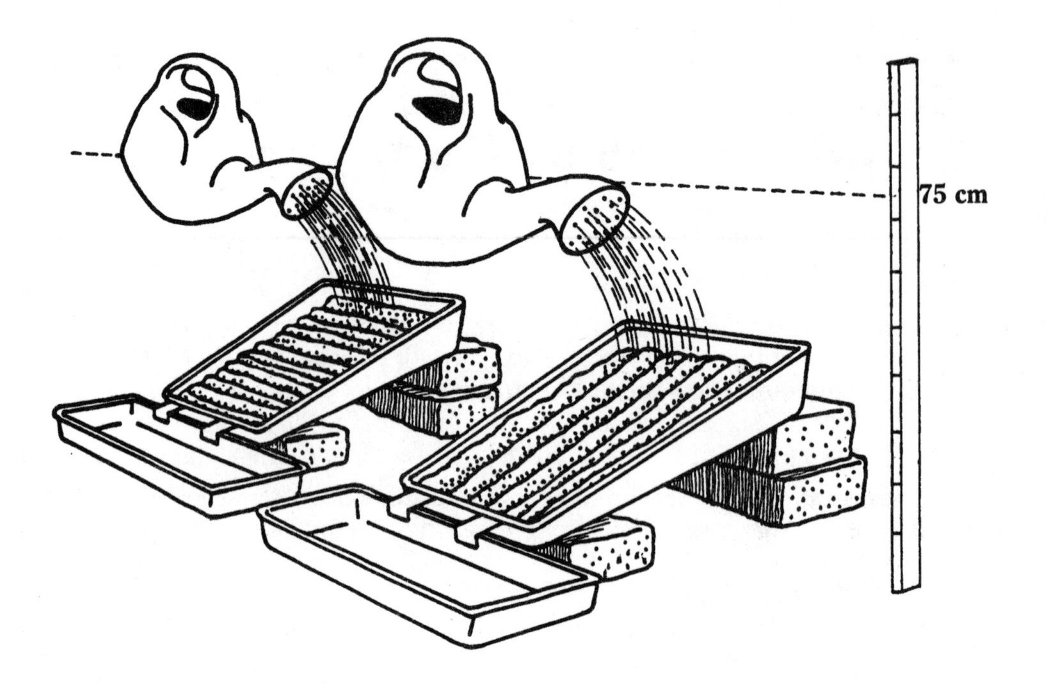

5. Fill the watering can. Simulate a rainstorm on each model hillside by sprinkling approximately 2 L of water on the upper end of each hillside from a height of 75 cm.

Describe how the water moves across the soil for the two different model hillsides. Which suffered more erosion? How can you tell?

____

____

____

6. Describe how a farmer in hilly country should plow the land to reduce the amount of erosion that occurs when it rains.

____

____

7. Can you think of other ways that a farmer could reduce soil loss from the fields?

____

____

GUIDE TO ACTIVITY 15

How Can Farmers Reduce Erosion Caused by Rain?

◆What is happening?

This activity simulates a technique used by many farmers to reduce erosion in plowed fields—**contour plowing.** In contour farming, one of the easiest and most widely accepted soil conservation practices, plowing is done *across* the slope of the land—that is, *on the contour*. When farmers plow on the contour in a hilly field, instead of plowing along the usual straight field boundaries and straight rows, they follow curved lines wherever necessary to stay at the same elevation.

Contour plowing alone will not stop erosion; the steepness and length of slope of fields affect the rate of erosion, as does the type of crop being grown and the condition of the soil. However, contour plowing can reduce soil erosion by as much as 50% on a wide range of soil and slope conditions. Contour farming is most effective at reducing erosion if it is combined with such practices as crop rotation and returning organic matter to the soil.

Vocabulary

- **Contour plowing:** Plowing done in accordance with the natural outline or shape of the land by keeping the furrows or ditches at the same elevation as much as possible. Contour plowing reduces runoff and erosion.

◆Time management

Assuming the students have done Activity 14, this activity should only take about one class period to complete.

◆Preparation

All of the topsoil used in this activity should be of a similar type. If the soils are different, it will be difficult to know whether the differences in erosion are caused by differences in the soil or by the different directions of the ditches on the two model hillsides.

Any similar-sized tray or baking pan can be substituted for the paint roller tray or the collecting basin.

◆Suggestions for further study

Both this activity and Activity 14 use the same apparatus. If you do not have enough watering cans, you can make a sprinkler by using a small nail to punch holes in the bottom of a coffee can or other large can. The plastic top from the coffee can can be used to stop the flow of water. You can also use a plastic soda or milk container with a cap. Punch holes near the top on one side and invert to use after the cap is replaced. You may wish to ask each group to provide its own watering device.

Repeat the experiment afte changing the slope by adding or subtracting supports. Be sure to use the same amount of water added at the same rate each time. How does slope affect erosion rate?

◆Answers

5. On the model hillside with ditches going straight down the slope, the water runs downhill very quickly, carrying a great deal of soil. As the "rain" continues to fall, the ditches get deeper and wider.

The model hillside with ditches running across the slope suffers very little erosion. The upper ditches act as dams, holding back and slowing down the water, allowing much of it to sink into the soil.

6. In hilly country, plowing so that the ditches go across the slopes of the hillsides rather than straight up and down the hills can greatly reduce soil loss.

7. Mulching helps to reduce soil loss. After harvesting grain crops, farmers sometimes distribute the straw from the stems of the plants on the fields to serve as mulch. Planting grass or some other type of ground cover on fields after crops have been harvested also helps to prevent soil from washing away or being carried away by the wind.

ACTIVITY 16 WORKSHEET

How the Soil Affects Acid Rain

◆Background

Scientists measure the acidity of solutions on the pH scale ranging from 0 to 14. A pH of 7 is neutral—neither acidic nor basic. A pH greater than 7 indicates that the solution is basic (alkaline); pH readings lower than 7 indicate that the solution is acidic.

Acid rain (rain having a pH of 5 or lower) is damaging plant and animal life in the United States, Canada, and Europe. Forests are dying; some streams and lakes have become so acidic that fish can no longer live in them. However, other areas receiving acid rain seem to have suffered little obvious damage. The composition of the soil may be one of the factors that determines the severity of the environmental effects of acid rain.

In this activity silica sand and silica sand with limestone will serve as models of actual soil in our study of acid rain.

Materials

Each group will need

- safety goggles for each person
- wide-range pH paper (either 0–14 or 1–13) and its color chart
- 100 ml graduated cylinder
- two 250 ml cups
- two 150 ml beakers (or four 5 oz paper cups)
- 200 ml of "acid rain" solution
- 100 ml of clean silica sand (pure white quartz sand)
- 100 ml of clean silica sand and limestone (Tums™)
- scissors
- rags, sponges, or towels to clean up spills
- a water source
- a waterproof marker
- distilled water (optional)
- soda, lemon juice, vinegar

◆Objective

To illustrate how certain types of soil can neutralize acid rain

◆Procedure

1. Determine the pH of some common acids—soda, lemon juice, and vinegar—using pH test paper as follows:

Tear off a short strip of pH paper (4 to 5 cm long) from the roll. If you are using pH paper that comes in the form of individual test strips, use the entire strip—do not tear or cut it.

Dip the strip into one of the substances (use a different piece of pH paper for each substance). Thoroughly wet the bottom 2 cm of the strip. Allow any excess liquid to drip back into the container.

Hold the wet end of the strip beside the color chart that accompanies the pH paper. Find the color on the chart that most closely matches the color of the wet strip.

Repeat for the remaining two substances.

Record your measurements at right:

Discard the strip—pH paper cannot be reused.

pH of common acids

pH of soda = ________

pH of lemon juice = ________

pH of vinegar = ________

2. Measure out 100 ml of clean sand (pure white quartz sand) using the graduated cylinder. Label one of the 250 ml cups "plain sand." Place 100 ml of clean silica sand in the cup. Label the second 250 ml cup "sand and limestone." Place 100 ml of the sand containing limestone dust in that cup.

3. Determine the pH of the "acid rain" solution using pH test paper as you did in step 1.

Record the pH of the acid rain solution:

pH of the acid rain solution = ________ .

! Safety procedures for working with acids and bases: *Wear goggles* whenever you are working with solutions of unknown pH.

Do not attempt to measure the pH of strong acids (such as battery acid) or strong bases (such as Drano™ or other drain cleaning compounds).

If you get a solution on your skin, quickly wash it off with a large amount of water. Tell your teacher about the accident.

Never eat or drink in a laboratory.

4. Measure out 100 ml of acid rain solution using the graduated cylinder. Pour 100 ml of the acid rain solution into each of two 150 ml beakers. Add 100 ml of this solution to the cup containing the plain sand. Add 100 ml of this solution to the cup containing the sand and limestone. Allow the acid rain solution to soak into the model soils for five minutes.

5. While holding the cup over a 150-ml beaker, stick the blade of the scissors through the bottom of one of the cups containing the sand and acid rain solution, producing a small narrow slit through which liquid can escape.

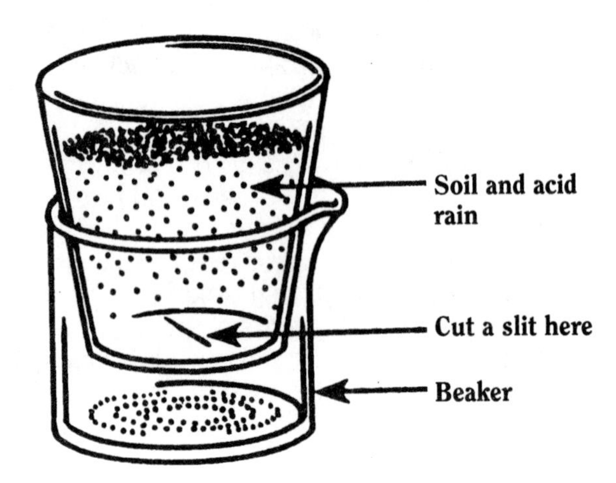

Set or hold the cup above a clean beaker to collect the liquid dripping from the cup. When the cup has stopped dripping, measure the pH of the solution and record your results in the data table below.

Repeat the procedure for the cup containing the sand and limestone and acid rain.

pH of acid rain absorbed by different soils

Type of soil	Color of strip	pH
plain sand	__________	__________
sand and limestone	__________	__________

6. Compare the pH that you recorded in step 3 for the acid rain solution with the pH of the solution that passed through the plain sand. Circle the letter that most accurately completes the following statement:

The pH of the solution passing through the plain sand

a. stayed about the same.

b. increased (approached pH 7, a neutral pH).

c. decreased (approached pH 1—became more acidic).

7. Compare the pH that you recorded in step 3 for the acid rain solution with the pH of the solution that passed through the sand and limestone. Circle the letter that most accurately completes the following statement:

The pH of the solution passing through the sand and limestone

a. stayed about the same.

b. increased (approached pH 7, a neutral pH).

c. decreased (approached pH 1—became more acidic).

8. Which type of area do you think would be most likely to be damaged by acid rain—one where the soil is sandy and contains little limestone, or an area with limestone-rich soil? Explain why you believe this to be the case.

____

____

____

9. Using the knowledge you gained from this activity, suggest a method to help reduce the problems of acid rain.

____

____

____

____

____

____

____

____

____

GUIDE TO ACTIVITY 16

How the Soil Affects Acid Rain

Vocabulary

- **Acid deposition:** Commonly called acid rain. It is water that falls to or condenses on the Earth's surface as rain, drizzle, snow, sleet, hail, dew, frost, or fog with a pH of less than 5.6.

◆What is happening?

There are many kinds of acid precipitation: rain, snow, sleet, hail, fog, dew, and frost. The precipitation in some areas of the northeastern United States often has a pH between 4 and 5—sufficiently acidic to injure or kill many types of living organisms. Although the popular press often uses the term acid rain in a generic sense to describe this problem, students should learn that there are multiple sources of acid deposition in the environment. Technically, dew and frost are not precipitation, so **acid deposition** is the most comprehensive term.

Scientists describe the acidity of liquids in terms of their pH. The pH scale ranges from 0 (strongly acidic) to 14 (strongly basic). Pure distilled water is neutral, and has a pH of 7. Each unit on the pH scale indicates a factor of 10 change in acidity. A solution of pH 6 is ten times as acidic as a pH 7 solution; a solution of pH 5 is one hundred times as acidic as a pH 7 solution.

Lakes with a pH of about 6.5 (just slightly acidic) provide optimal conditions for maintaining healthy populations of fish and other aquatic organisms. But as a result of acid precipitation, the lakes in certain areas of the country are becoming more and more acidic. When the pH of a lake's water falls below 5, very few organisms can survive.

Lakes and streams that are surrounded by soils and rocks that contain limestone and other naturally occurring basic substances may not be harmed immediately by acid precipitation. The calcium carbonate in limestone neutralizes some of the acid, preventing the pH of the water in the lakes from falling to dangerously low levels when it rains.

However, there are many places where the soil is thin and the main type of rock present is granite. Granite breaks down into silica and clays that lack the ability to neutralize acids to any large degree. In some areas where granite (and granite-based soil) is the main geologic feature, the lakes have become so acidified that they can no longer support plant and animal life. Forests surrounding these lakes often exhibit signs of damage from acid rain, too.

◆Time management

Due to the newness of materials and procedures, this activity should take about two class periods to complete and clean up.

◆Preparation

Making the acid rain solution:

Mix 100 ml of vinegar with 900 ml of water. This solution should have a pH of approximately 4. Use standard white vinegar (5% acetic acid), available in most grocery stores.

Tap water is acceptable for mixing the solution. However, if the tap water in your area is very acidic or basic, you may have to use distilled water for mixing the solution, or change the ratio of vinegar to water. To raise the pH when the solution has a pH of less than 4, add additional water; if the pH is higher than 4, add additional vinegar. Each group performing the activity will require 200 ml of this acid rain solution.

Preparing the sand:

Any clean white silica (quartz) sand may be used for this activity. This type of sand is sold in garden shops and department stores as play sand for children's sandboxes. (To check if the sand is silica: The grains will scratch glass if it is silica sand—all other sands will not.) If the sand is extremely dry, moisten it slightly with tap water before use. It should feel slightly damp to the touch, but not be dripping water.

Making the sand and limestone mixture:
Pure limestone is composed mainly of calcium carbonate ($CaC0_3$), as is Tums™, the over-the-counter stomach acid remedy. Crush 25 Tums™ into a fine powder. Thoroughly mix this powder with 500 ml of sand. Each group performing the activity will need 100 ml of this mixture.

Pulverized or powdered limestone (calcite) or marble chips from a garden shop will be cheaper than Tums™, but harder to crush. Any standard mortar and pestle should crush limestone—or just use a hammer.

Working with acids and bases:
You can use this activity to teach students about good laboratory safety techniques for working with acidic and basic materials. Although the dilute vinegar used as the acid rain for this activity is no more dangerous than most undiluted salad dressings, it should be treated with the same respect given to stronger acids and bases. The most important rules to stress to students are:

1. Wear eye protection (goggles or safety glasses) during laboratory activities.
2. Never eat or drink in a laboratory.
3. Never test the pH of strong acids (such as battery acid) or strong bases (such as drain-cleaning compounds).

Using pH Paper:
Be sure that the pH paper that you are using is fresh. These papers can give incorrect results if they have gotten damp or been exposed to the air for too long. If possible, check the pH paper's accuracy using a buffered solution of known pH or against a pH meter.

The standard pH testing papers come in two forms, "wide range" and "short range"; use the wide range type for this activity. Wide-range papers will indicate an approximate pH of liquids from about pH 1 to pH 12. Short-range papers are useful for making more precise determinations of pH within a narrow range of pH values (for example, between pH 3.4 and 4.8).

The exact pH range of the paper and the colors indicating a particular pH value will vary for different brands of pH papers. Be sure that the color comparison chart that you are using is the correct one for the pH paper that you are using.

! First aid in the event of an accident: Whenever an acid or base spills on you or your clothes, immediately wash the affected area with large volumes of running water. Consult the school nurse or a physician as soon as possible. If a caustic liquid splashes into your eyes, rinse your eyes under running water for at least 15 minutes. While you are rinsing your eyes, have someone else seek medical help for you.

◆Suggestions for further study

Milk of magnesia or baking soda are often taken as home remedies for excess stomach acid. Your stomach secretions have a pH of about 2. See how much milk of magnesia or baking soda are required to neutralize 100 ml of vinegar or lemon juice—liquids that also have a pH of about 2. The acidic solution has been neutralized when its pH reaches 7.

Collect some precipitation from your area and measure its pH. You could do this a number of times over a period of time and graph the results.

Discuss the sources of the materials causing any local acid deposition.

◆Answers

The following pH values are intended only as guidelines. Your results may differ by ±1 pH unit.

2. The acid rain solution should have a starting pH of about 4.

Sample data

pH of acid rain absorbed by different soils

Type of soil	Color of strip	pH
plain sand	*Varies**	*4*
sand and limestone	*Varies**	*6*

*The color corresponding to a particular pH depends on the brand of pH paper being used.

6. The pH of the solution passing through the plain sand: a. stayed about the same.

7. The pH of the solution passing through the sand and limestone: b. increased (approached pH 7, a neutral pH).

8. Soils containing little limestone are most likely to be damaged by acid rain. Such soils have very little ability to neutralize any acid precipitation striking them.

9. Adding limestone to soils and lakes could help reduce the problems caused by acid rain.

MODULE 4

Water, Water, Everywhere

◆Introduction

• Water in its liquid state covers more than 70% of our planet.

• If all the fresh water on Earth were divided equally among the five billion people alive today, each person would get about 105 million L of water.

• Human beings require about 2.4 L of water each day to sustain life, yet each year drought kills thousands of people and makes refugees of hundreds of thousands more.

Water, unlike minerals and fuels, is a renewable resource. However, securing adequate supplies of good-quality water for drinking, washing, and manufacturing purposes is becoming increasingly difficult as Earth's human population grows past the five billion mark. The popular press may someday chronicle a water crisis that causes worse economic and social problems than the energy crisis that began during the 1970s.

This module consists of activities that will challenge you to purify water, decide how to conserve a family's limited water supply, and investigate the amount of water contained in living tissues. You will gain hands-on experience in using metric units for measuring volumes of water. You will also learn about the amount, location, and quality of water on Earth; the water cycle; and how our jobs depend on water.

SUMMARY OF MODULE 4

Water, Water, Everywhere

◆Instructional Objectives:

After completing the activities and readings for Module 4, you should be able to

- describe the relative proportions of Earth's water existing as subsurface water, ice, lakes and streams, ocean water, and water in the atmosphere [Activity 17]
- describe how much water an average family of four uses in the United States each day [Activity 18]
- determine how much water leaks from a faulty plumbing fixture per hour, per day, and per year [Activity 19]
- measure the water contained in living tissues [Activity 20]
- demonstrate the procedures that municipal water plants use in order to purify water for drinking [Activity 21]
- describe the Earth's water cycle [Reading 13]

◆Preparation

Study the following readings for Module 4:

Reading 13: Conservation and the Water Cycle

Reading 14: Where Can People Get the Water They Need?

Reading 15: Water Treatment

Reading 16: Practical Tips for Conserving Water

◆Activities

This module includes the following activities:

Activity 17: All the Water in the World

Activity 18: We'll Form a Bucket Brigade

Activity 19: Take Me to Your Lost Liter

Activity 20: The Incredible Shrinking Head

Activity 21: Taking the Swamp Out of "Swampwater"

ACTIVITY 17 WORKSHEET

All the Water in the World

◆Background

This activity asks, "How much water is there on Earth, and where is it located?" You will use 100 L of water to represent "all the water in the world." You will work with a group to guess how to divide this water among different containers that represent the categories of water listed below. Do not worry or be embarrassed if your guesses are inaccurate.

◆Objective

To describe the relative proportions of Earth's water existing as groundwater, ice, lakes and streams, ocean water, and water in the atmosphere

Water is found in the following different forms and locations:

- **Groundwater** is contained in soil and rocks. Some of this ground water is salty.
- Oceans, rivers, lakes, and salt lakes are **surface water.**
- The atmosphere contains **water vapor** (water in the form of a gas).
- The rest of Earth's water is in the form of ice, such as glaciers and **sea ice** (ice formed by freezing of ocean water).

◆Procedure

1. The note cards should be labeled as follows:

- all the water in the world
- oceans
- freshwater lakes
- saltwater lakes
- rivers
- groundwater
- sea ice and glaciers
- water vapor

2. There are 100 L, or 100,000 ml of water in the "all the water in the world" container. Since each milk jug used to fill this container holds 3.8 L, it took a little more than 26 jugs to get 100 L (3.8 L/jug x 26.3 jugs = 100 L).

3. If 100 L is "all the water in the world," how many liters of water would be in the oceans?

Discuss this with other members of the group and decide together on your best guess as to how much water this would be. Record your best guess below:

_________ L of water (out of 100 L total) are in the oceans.

4. Using your estimate for the amount of water in the oceans, calculate how much water would not be in the oceans as follows:

100 L (total) - ________ L (in the oceans) = ________ L of water not in the oceans.

Materials

Each group will need

- several empty 3.8 L (1 gallon) plastic milk jugs
- one or more of each of the following sizes of graduated cylinders: 1000 ml, 100 ml, and 10 ml
- other containers for dividing up "all the water in the world" (2 or 3 buckets or aquaria; 6 or 8 beakers or cups [several each of a range of sizes—1000, 500, 250, and 100 ml beakers]; 6 or 8 test tubes and test tube holders)
- 10 note cards to serve as labels for the containers
- masking tape
- a mop and sponges or rags to clean up spills

Vocabulary

- **Groundwater:** Water below the water table in the zone of saturation. Groundwater fills the spaces between soil and sediments or lies in the cracks and crevices in rocks.
- **Sea ice:** Solid water that forms by the freezing of ocean or sea water. Sea water has an average freezing point of -1.9° C (28.6° F). However, the freezing point varies with the salinity of the water. Sea ice is much of the area around the Earth's North Pole.
- **Surface water:** All water, fresh and salty, on the Earth's surface. Oceans, lakes, streams, snow, and glaciers are all surface water.
- **Water vapor:** The gaseous state of water.

5. When your class agrees on how much water is not in the oceans, remove that amount of water from the original 100 L of water.

Save this water that you remove from the "all the water in the world" container. Your next task is to decide how to divide this "not in the oceans" water among the other categories of water.

In the spaces at left, write the group estimates of the volume of water that should be placed in each category.

Volume of water

freshwater lakes: ______

saltwater lakes: ______

rivers: ______

groundwater: ______

glaciers and other ice: ______

water vapor: ______

Note: Express these volumes in terms of milliliters if that makes the task easier. Remember, 1 L = 1000 ml

6. Label an appropriate-sized container for each of the remaining categories of water. From the bucket containing the "not in the oceans" water, measure out the amount of water that you assigned to each of these categories. Pour the water that you measure out into its labeled container.

The water remaining in the "all the water in the world" container should be what you have estimated as the amount of water in the oceans. At this time you should change the label from "all the water in the world" to "oceans."

7. You should now have finished the process of dividing 100 L of water representing "all the water in the world" into 7 major categories on the basis of your previous knowledge.

Your teacher will now demonstrate the correct relative amounts of water in each category.

Were your estimates for each category close to the actual amount of water in the category? Were there any major surprises? Describe your findings below:

____

____

____

____

____

____

____

____

____

GUIDE TO ACTIVITY 17

All the Water in the World

◆What is happening?

When asked to estimate the water supply on Earth, most people overestimate the fresh water available in streams, rivers, and lakes, and greatly underestimate the amount of water in underground deposits and in glaciers.

Very few people will guess that the total amount of fresh water that is reasonably accessible for human use (the water in streams, freshwater lakes, the atmosphere, or less than 1000 m underground) is less than 0.33% of the total amount of water on Earth. Assuming that 100 L constitutes "all the water in the world," Earth's reasonably accessible fresh water would have a maximum volume of about 325 ml—less than the amount required to fill a soft drink can.

In practical terms, the actual amount of accessible fresh water is much smaller, because much of this reasonably accessible water is poor-quality groundwater. Even where groundwater is plentiful and of excellent quality, it is not possible to extract more than a small fraction of the water in an aquifer.

Reducing all the water in the world to 100 L placed in a trash can makes the problem of dividing up the different types of water resources more concrete, and also demonstrates graphically the weight and volume of 100 L of water.

Even when you know the relative amounts of water in different locations, it is very difficult to visualize. Reference materials often describe the Earth's water supply in terms of cubic miles of seawater, or acre feet used for irrigation, or percent of the total water available. The numbers used in these types of descriptions are large and abstract.

◆Time management

Because of the necessary student/teacher interaction, this activity should take about two class periods.

Allow the students to work in small groups to decide how to divide up "all the water in the world." After all the groups have decided upon the relative proportions of the various water resources, write the answers of all the groups on the blackboard or overhead projector. Allow the class to discuss their differences and if possible reach a class consensus on the division of the water resources.

Since they both require the use of 100 L of water, you may wish to perform both this activity and Activity 18 outside on the same day—this way you can use the same 100 L of water for both activities.

A water source such as a garden hose is very handy when setting up these activities.

◆Preparation

Any type of container can be used to hold the different categories of water. If you do not have enough beakers, you can use bottles or cups (plastic, if possible). If you lack graduated cylinders, you can use measuring cups to measure the large volumes of water and medicine dispensers to measure the smaller volumes.

Before beginning this activity with your class, measure out 100 L of water to represent all the water in the world. This is equivalent to 26.3 gallon milk jugs full of water. The water may be placed in a 122 L (32 gallon) trash can, or three 38 L (10 gallon) aquaria. You may wish to add some food coloring to the water to make the water easier to see. You can use the same 100 L of water for each class performing this activity. Be sure to have a mop and plenty of rags or sponges for cleaning up.

You may want to have just one group of students measure and pour the water for the demonstration rather than having each group measure and pour.

After the discussion, have students actually transfer the water from the "all the water" container into the other labeled containers before showing them the correct answers (given below).

You may wish to have a separate set of containers for the "class answer" and the "correct answer," so that students can compare the differences in the volumes.

◆Suggestions for further study

Have students contact the local water utility to find out about how water is used in their community. Also, where does this water come from—wells, streams, collectors? Is the supply ample? How much treatment is necessary to make the water safe for use?

◆Answers

The relative amounts and location of fresh, salty, and frozen water on the planet Earth are shown in the table that follows.

The final column of the table gives the volume (in ml) for each of these categories of water, assuming that 100 L of water is "all the water in the world."

Category	% of the total water in world	Equivalent in ml (out of 100 L "all the water in the world" container)
freshwater lakes	0.009	9.0 ml
saltwater lakes	0.008	8.0 ml
rivers	0.0001	0.1 ml
groundwater water	0.625	625.0 ml
sea ice and glaciers	2.15	2,150.0 ml
atmospheric water vapor	0.001	1.0 ml
all oceans	97.2	97,206.9 ml

These figures are adapted from information contained in the U.S. Government Pamphlet 1977-0-240-966/44 "Water of the World," by Raymond Nace.

ACTIVITY 18 WORKSHEET

We'll Form a Bucket Brigade

◆Background

Human beings require a minimum of 2.4 L (about 2/3 gallon) of water per day to sustain life. However, the average American uses about 100 times more water than this every day at home. An average family of four in the United States might use about 900 L of water per day for the following purposes:

Approximate daily water use by a family of four in the U.S.

Use	Liters per day
Drinking and cooking	30
Dishwasher (3 loads per day)	57
Toilet (16 flushes per day)	363
Bathing (4 baths or showers per day)	303
Laundering clothes	130
Watering houseplants	4
Rinsing garbage into disposal unit	13
Total daily use:	**900 L**

(A reminder: 1 gallon = 3.8 L; 26.3 gallons = 100 L.
The total daily water use of 900 L is equal to about 237 gallons.)

Materials

Each group will need

- a schoolyard or parking lot with a water source
- two 122 L (32 gallon) trash cans
- empty milk jugs and/or buckets (as many as possible)
- 100 L of water (If possible, use the same water that you used for the "All the Water in the World" activity.)
- a watch or clock with a second hand
- a meter stick (optional)

Note: *You can do this activity inside a building, but try to choose a place where some spilled water will not be a problem.*

◆Objective

This activity provides a real-life model for how much water a family typically uses, allowing participants to experience firsthand how much effort is required to transport water, and shows that when people really want to, they can sharply reduce their water usage.

The story begins:

One cold January, the Smith family rent a house in the mountains for a ski vacation. The house, though old, has all the comforts of home—three bathrooms, a complete laundry room, dishwasher, and garbage disposal, plus a newly installed solar hot water heating system. Unfortunately, the weather gets so cold one night that a water main in town breaks, and the Smiths find out that the house will have no water service from the local utility for the entire week. What should they do—go back home or try to find another water supply?

Mr. Smith learns from a neighbor that there is an unfrozen spring 100 m from the house that could still be used for drinking water. Mrs. Smith, who is a mechanical engineer, discovers that if the municipal water line coming into the house were shut off, the water in the storage tank for the solar water heater could be routed directly into the plumbing system. The water system in the house will work as long as the storage tank is kept filled with water from the spring.

Mr. and Mrs. Smith discuss the decision with their two children: Alice, 14, and Sam, 12. The family decided to form a "family bucket brigade" from the spring to the house, fill the storage tank each day, and continue their vacation. The storage tank can hold about 900 L of water.

Your role:

For this part of the activity, you will pretend to be one of the Smiths, who need to carry 900 L of water 100 m to the house.

◆Procedure

1. Place the two trash cans 100 m apart. (You can measure the distance with a meter stick or estimate it—100 m is about 150 paces for an average-size person.) If you cannot go 100 m in a straight line on the schoolyard, set up a curving course.

2. Place 100 L of water in one of the trash cans. This will be the "spring."

3. Select four people to start out as the Smith family, equip each person with as many buckets and milk jugs as he or she can carry, and have them transfer the 100 L of water to the house (represented by the other trash can) 100 m away.

Record the time when the Smiths begin carrying the water.

__________ am / pm

Record the time when the Smiths finish moving the first 100 L of water from the spring to the house.

__________ am / pm

How long did it take them? __________ minutes

4. The Smiths may feel a little tired, but so far they have only carried 11% of the water required to fill the tank. They still have 800 L to go. To save your water (since this is role playing), have the Smiths bring the same 100 L back from the house to the spring rather than getting additional water out of the faucet you are using.

5. The Smiths should continue carrying the water back and forth until the 100 L of water has changed cans a total of nine times, and the Smiths have carried the equivalent of 900 L of water 100 m to the house.

Record the time when the Smiths finish moving the entire 900 L of water from the spring to the house.

__________ am / pm

How long did it take them? __________ minutes

The story continues:
After carrying all that water, the Smiths are too tired to ski very much. They come home early, have spaghetti for lunch, wash the dishes, and launder their bucket brigade clothes (which got muddy at the spring). After eating dinner, washing more dishes and clothes, watering the houseplants, and taking long, hot showers, they go to bed.

It is snowing too hard the next day to ski, so they stay in the house all day. When he tries to start the dishwasher after lunch, Mr. Smith discovers that they are out of water! Sam and Alice groan and say that they would rather be grounded until they are 21 than carry 900 L of water to the house every day. They point out that they haven't even been in the house a full 24 hours since carrying the water.

Your assignment:
Discuss water conservation measures with your group. Suggest five different ways that the Smiths can save water (and by doing so, save the ski vacation).

(1) ____________________

(2) ____________________

(3) ____________________

(4) ____________________

(5) ____________________

Defend your suggestions:
Do the suggestions that your group decided upon really reduce the total amount of water that the Smiths are using each day, or do the suggestions simply change the location where they use the water?

For example, if Sam does not shower all week, that reduces the total water that he uses. However, if he showers at the ski lodge, he has simply changed the location of his water usage. That way, he takes less water from the tank in the house, but uses the same amount of water overall each day.

____

____

____

____

____

____

____

____

____

____

____

GUIDE TO ACTIVITY 18

We'll Form a Bucket Brigade

◆What is happening?

Americans are dependent on the convenience of modern plumbing. Water is so effortlessly obtainable from a tap that we have forgotten that it is heavy and hard to transport by hand. Hopefully this activity will help students become more conscious of their own water use habits.

In most places in the United States (in sharp contrast to much of the rest of the world), water is cheap and readily available. U.S. citizens take our water sources for granted, and have come to consider unlimited water as a right. Yet data from the United Nations Development Program indicate that roughly half of the world's population needs to carry water supplies. As our population continues to increase, and as many people settle in drier areas, we will need to make strenuous efforts at personal, agricultural, and industrial water conservation.

The water use figures for this activity are adapted from *How Much Water in a 12 Ounce Can? A Perspective on Water-Use Information* by I.C. James, J.C. Kammerer, and C. R. Murray. USGS Annual Report, Fiscal Year 1976, U.S. Government Printing Office: 1980-311-348/56. More recent sources provide somewhat larger estimates; for example, 340 L per person per day, or a total of 1360 L per day for a family of four.

◆Time management

This activity should take about two full class periods, and could take up to three.

The bucket brigade of four people will require two round-trips to the spring to move 100 L of water. At that rate of transfer, it will take about half an hour to move 900 L of water. (**Note:** If you cannot spend this much time on a single activity, you may wish to stop the bucket brigade after one or two transfers of 100 L. The point of this activity is to show that while 100 m is not very far, moving a typical family's water supply even a short distance requires a lot of work.)

To motivate and speed up students, make this a timed relay race with a prize for the fastest bucket brigade. Do not allow students to run while carrying the milk jugs, however—restrict them to a fast walk. Allow students at the spring to have the jugs full and ready to go when the bucket brigade members arrive.

Since they both require the use of 100 L of water, you may wish to perform both this activity and Activity 18 outside on the same day—this way you can use the same 100 L of water for both activities.

A water source such as a garden hose is very handy when setting up these activities.

◆Preparation

Use large trash cans to hold the 100 L of water. They are much safer to use than glass aquaria, which are easily broken. Perform this activity outside in the schoolyard if at all possible. This will reduce cleanup hassles. Use only one spring and one house for the entire class, but let everyone who is physically able participate in the bucket brigade. Many middle school students can safely carry 12 to 15 L of water (3 or 4 gallons, weighing roughly 20 to 30 pounds) per trip.

◆Suggestions for further study

Discuss with your class how a permanent water shortage in your region might change their lives. If the water system in their homes suddenly stopped working and could never be fixed:

- What immediate changes would you and your family have to make in your way of life?
- What are some of the things that you could no longer do?
- Would certain appliances in your home become useless?
- Would a water shortage affect your parents' jobs? Would they still have jobs?
- What would happen to people who build washing machines or other water-using machines, or who work as plumbers?
- How would the economy of the town be affected?
- How would the town deal with emergencies like fires?
- What would happen to farmers in the area?
- What would happen to industries in the area?

◆Answers

3. Times will vary, but a motivated bucket brigade of four people can probably move 100 L by making two round trips between the house and the spring. Allowing time to dump the water into the house, this should take about three minutes.

5. Moving 900 L at this rate will require about 27 minutes.

Conservation measures may include:

- washing clothes less frequently
- running the dishwasher once a day, rather than after every meal
- fixing any leaky plumbing fixtures
- not flushing the toilets after every use
- reducing the amount of water the toilet uses per flush
- using a special shower head to reduce water use
- taking quick showers

All of the above suggestions would actually decrease the amount of water being used daily by the Smiths. Showering in the ski lodge or washing clothes at a commercial laundry would reduce the use of water at the house, but would not reduce the overall water consumption of the Smith family.

ACTIVITY 19 WORKSHEET

Take Me to Your Lost Liter

Materials

Each group will need

- a container for catching 1 L of water (1 L = 1000 ml)
- a watch or clock with a second hand
- a source of dripping water (Use a sink, or you can make multiple "leaky water sources" by poking a small hole in a gallon milk jug.)
- a container to catch the excess dripping water such as a bucket, cut-off bottom of a milk jug, or a pie pan

Observations:
Drops in 10 seconds

Trial 1: _______ drops

Trial 2: _______ drops

Trial 3: _______ drops

Average of the three trials =

_______ drops in 10 sec

Average drops in 10 sec

_______ ÷ 10 sec =

______ drops/sec.

◆Background

It's easy to lose a liter (of water) at home. Faucets wear out and begin to drip; everyone occasionally leaves the water dripping a little after using a sink. Over time, small leaks like these add up to large amounts of wasted water.

◆Objective

To determine how much water leaks from a faulty plumbing fixture per hour, per day, and per year

◆Procedure

1. Start your water source dripping at a good, steady rate. Be sure that you are getting individual drops, rather than a steady stream of water.

2. Count how many drops fall during a 10-second interval. Record your observations in the table at right. Leaving the water dripping, repeat this measurement three times, and average the three readings. Use the average to calculate the number of drops falling in one second.

3. See how long it takes your water source to drip 1 L. First, set up your 1 L container so that it will catch the dripping water.

Record the time when you place the container under the dripping water.

starting time = __________ am / pm

4. Record the time when 1 L has dripped into the container.

finishing time = __________ am / pm

5. How much time did it take for 1 L to drip into the container?

__________ seconds / minutes

6. At that rate of dripping:

How much water would be wasted in one hour? _______ L

How much water would be wasted in one day? _______ L

7. Human beings need to drink about 2.5 L of water each day to survive. How many people could survive by drinking the water that would be wasted by your water source in one day?

______ people

8. Assume that your water source will continue dripping at that same rate for one year.

How much water would be wasted in one year? _______ L

9. If 1000 homes in your community have water leaks that waste water at this rate, how much water does the entire community waste each year?

_______ L

GUIDE TO ACTIVITY 19

Take Me to Your Lost Liter

◆What is happening?

Leaky plumbing fixtures waste enormous amounts of water and energy each year in the U.S. This activity focuses on just how much water might be wasted by a single leak in a single home. Extrapolating the observations to the community shows how large the problem really is.

◆Time management

Collecting 1 L of water for this activity may take anywhere from 5 to 50 minutes, depending on the rate of dripping. You may wish to begin Activity 20, "The Incredible Shrinking Head," or assign one of the readings from Module 4 while the water is filling the container. Allow one-and-a-half to two periods to complete the whole procedure.

◆Preparation

Students can perform this activity as a homework assignment if you spend part of a class session helping them calibrate a container for catching 1 L of water. Milk jugs and soft drink containers are excellent for this purpose. To calibrate them, pour an accurately-measured liter of water into the container. Use a waterproof marker to label the 1 L level on the container.

◆Suggestions for further study

Have your students check their homes for water leaks and determine how much water they can save in a year by stopping these leaks.

Collect water from a dripping faucet for one hour and multiply by 24, then by 365 to see how much would be lost in a year.

Find out how much water costs in your town. How much would 11,942,800 L cost? Who would pay for this wasted water?

◆Answers

Note: The following set of observations is included as a general guideline only. The actual results that students will obtain depend on the speed of dripping and size of drops from the water source they use.

Average of the three trials = *14.67* drops.

Calculate the average number of drops per second as follows:

14.67 (drops in 10 sec) ÷ 10 sec = *1.5* drops per second.

Questions 5–9: Answers for these questions depend on the rate of dripping that the students determine experimentally in steps 3 and 4. The following answers are included only as guidelines.

5. The faucet that we tested required *44* minutes to drip 1 L of water.

Sample data

Observations:
Drops in 10 seconds

Trial 1: *15* drops

Trial 2: *14* drops

Trial 3: *15* drops

Average of the three trials =

14.67 drops in 10 sec

Average drops in 10 sec

14.67 ÷ 10 sec =

1.5 drops/sec.

6. At that rate of dripping, *1.36* L would be wasted in one hour.

To obtain this figure, divide 60 minutes by the time required to waste 1 L.

60 min ÷ 44 min/L = 1.36 L wasted in an hour

At that rate of dripping, *32.72* L would be wasted in 1 day.

To obtain this figure, multiply the number of liters wasted per hour by 24 hours.

1.36 L wasted in an hour x 24 = 32.72 L wasted per day.

7. *13* people could survive by drinking the water that would be wasted by this water source in one day.

To obtain this figure, divide the number of liters wasted per day by 2.5 L, which is the daily water requirement of one person.

32.72 L wasted per day ÷ 2.5 = 13.088 (round off to 13)

8. In one year, about 11,942.80 L of water will be wasted if the drip continues at that rate.

To obtain this figure, multiply the number of liters wasted per day by 365.

32.72 L wasted per day x 365 = 11,942.80 L per year

(**Note:** Purists might multiply by 365 1/4 days per year to account for leap years. This gives a value of 11,950.98 L.)

9. At that rate, 1000 homes would waste 11,942,800 L of water each year.

ACTIVITY 20 WORKSHEET

The Incredible Shrinking Head

Materials

- an apple for the entire class to use as a control (object of comparison) for the experiment (it will remain uncarved)

Each group will need

- an apple
- a kitchen knife or strong plastic knife
- a scale or laboratory balance
- several sheets of newspaper or plastic sheeting

◆Background

Our bodies are about 70% water. We can survive for more than a month without food, but without water we would die in less than a week. The purpose of this activity is to find out what would happen if our skin suddenly lost its ability to hold water inside our bodies.

◆Objective

To measure the amount of water contained in living tissues

◆Procedure

1. Place your apple on a piece of newspaper. Using this diagram as a guide, carve a face on the apple by carefully cutting out slices for the nose, chin, and cheekbones. Make slits for the eyes and mouth.

2. When the carving is complete, weigh the "head" and record the weight in the data table below.

Set the head, along with an uncarved apple, in a safe, dry place. You will observe and weigh them once a week for the next three weeks. Notice how your apple changes in these three weeks compared to the uncarved apple.

The incredible shrinking head

Date	Weight (carved)	Weight (uncarved)	Observations
______	______ g.	______ g.	______
______	______ g.	______ g.	______

3. What changes in weight and appearance do you predict will happen to the head in the next few days?

______

______

4. Did the same types of changes occur in the apple that had not been carved? Why or why not?

______

______

5. Can you think of a way that you could carve a face in an apple but prevent it from shrinking?

____

____

6. Answer this question after observing the apple for two weeks:
When you eat a fresh apple, what percentage of the snack is solid apple, and what percentage is water?

____

____

7. Which apple lost more weight? Was your prediction correct?

____

____

GUIDE TO ACTIVITY 20

The Incredible Shrinking Head

◆What is happening?

All plants and animals living on land have evolved ways of preventing great changes in the amount of water contained in their bodies. The epidermis, which is the outer layer of the skin, prevents animals from drying out when they are in sunshine, or from gaining water and "swelling up" when rain falls on them. The cuticle, a waxy layer covering the surfaces of land plants, performs the same functions for plants.

In both plants and animals, when the outer layer of tissue is damaged, body water is lost. When you carve the apple, you remove over half of its water-conserving layer. The apple will shrink as it loses water to the air. If the air is humid, it will shrink slowly; if the air contains little water vapor, the apple will shrink very rapidly.

Most apple growers apply a thin coating of wax to their crop after harvest. The extra wax further reduces the amount of water lost through the apple's skin, which helps keep the apple fresh until it is sold. With proper storage, apples treated this way will remain fresh for a year.

◆Time management

One class period should be enough for the procedure (not counting the two week waiting period).

◆Preparation

Although not absolutely necessary, it is a good idea to leave one apple uncarved as a control for this activity.

Put the carved apples on a warm, sunny windowsill. The more air circulation there is around the apples, the quicker they will lose moisture, and the less likely they will become moldy.

◆Suggestions for further study

Art and science can go hand-in-hand in this activity. Encourage your students to be creative. Perhaps they would like to carve a different fruit or vegetable. Great—it shows that they are thinking about the concepts of the activity, and exploring other applications of the principle.

Have your students actually test some of their answers to question 5.

Conduct a discussion comparing the skins of apples and the skins of humans in relation to water loss.

◆Answers

3. Students may predict that the head will shrink, turn brown, or rot; accept any serious answer.

4. An uncarved apple will probably change very little in three weeks.

5. Answers will vary—freezing, coating the face with wax, putting it in a plastic wrapper, or other techniques may be suggested. (Note: You may wish to have your students test some of these techniques.)

6. After two weeks, the apple probably will have lost about half of its water, decreasing its weight by about 50%. Students may infer that all the lost weight is water that evaporated from the apple.

7. The carved apple lost more weight.

ACTIVITY 21 WORKSHEET

Taking the Swamp Out of "Swamp Water"

◆Background

Water in lakes, rivers, and swamps often contains impurities that make it look and smell bad; it may also contain bacteria and other organisms that can cause disease. In most places, you should not drink from surface-water sources until the water has been cleaned. This activity shows how water treatment plants turn "swamp water" into drinking water.

◆Objective

To demonstrate the procedures that municipal water plants use to purify water for drinking

◆Procedure

1. Using a beaker or graduated cylinder to measure, pour about 1.5 L of "swamp water" into a 2 L bottle. Describe its appearance and smell:

___

___

2. **Aeration** is the addition of air to water. It allows gases trapped in the water to escape and adds oxygen to the water. Begin aerating the swamp water as follows:

Put the cap on the bottle and shake the water vigorously for 30 seconds.

Continue the aeration process by pouring the water into either one of the cut-off bottles, then pouring the water back and forth between the cut-off bottles 10 times.

Does aeration change the appearance or smell of the water? Describe any changes that you observe.

___

___

3. Pour the aerated water into a bottle with its top cut off.

Coagulation is the process by which dirt and other suspended solid particles are chemically "stuck together" into **floc** so that they can be removed from water. Begin this process as follows:

With the tablespoon, add 20 g of alum crystals to the swamp water. Slowly stir the mixture for five minutes.

Materials

Each group will need

- 5 L of "swamp water"
- one 2 L plastic soft drink bottle with its cap (or a cork that fits tightly into the neck of the bottle)
- two 2 L plastic soft drink bottles—one bottle with the top removed, and one bottle with the bottom removed
- one 1.5 L (or larger) beaker or another soft drink bottle bottom
- 20 g of alum (potassium aluminum sulfate—approximately 2 tablespoons)
- fine sand (about 800 ml in volume)
- coarse sand (about 800 ml in volume)
- small pebbles (about 400 ml in volume)
- a large (500 ml or larger) beaker (or a large jar)
- a small (approximately 5 cm x 5 cm) piece of flexible nylon screen
- a tablespoon
- a rubber band
- clock with second hand or stopwatch

Vocabulary

- **Aeration:** The addition of air to water or to the pores in a soil.
- **Coagulation:** The process, such as in treatment of drinking water, by which dirt and other suspended particles become chemically "stuck together" (to form floc) so they can be removed from water.
- **Deposition:** The process of dropping or getting rid of sediments by an erosional agent such as a river or glacier; also called sedimentation.
- **Floc:** Clumps of impurities removed from water during the purification process; formed when alum is added to impure water.

4. Sedimentation is the process that occurs when gravity pulls the particles of floc to the bottom of the cylinder. This process is also called **deposition.**

Allow the water to stand undisturbed in the cylinder. Do not mix it or shake it. Observe the water at five minute intervals, and describe any changes in the water's appearance. Write your observations in the following spaces:

5 minutes: ____________________

10 minutes: ____________________

15 minutes: ____________________

20 minutes: ____________________

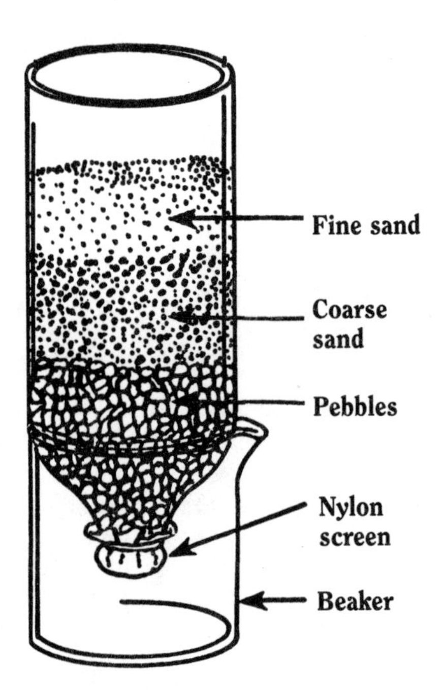

5. While you are making the above observations, construct a filter from the bottle with its bottom cut off as follows:

Attach the nylon screen to the outside of the neck of the bottle with a rubber band. Turn the bottle upside-down, and pour a layer of pebbles into the bottle—the screen will prevent the pebbles from falling out of the neck of the bottle.

Pour the coarse sand on top of the pebbles.

Pour the fine sand on top of the coarse sand.

Clean the filter by slowly and carefully pouring 5 (or more) L of clean tap water through it. Try not to disturb the top layer of sand as you pour the water onto it.

6. Filtration through a sand-and-pebble filter removes most of the impurities remaining in water after coagulation and sedimentation are completed.

After a large amount of sediment has settled on the bottom of the bottle of swamp water, carefully—without disturbing the sediment—pour the top two-thirds of the swamp water through your filter. Collect the filtered water in the beaker. Pour the remaining (one-third bottle) of swamp water into the class collection bucket.

Your treatment of the swamp water is now complete. Compare it with a sample of untreated swamp water. How has treatment changed its appearance and smell?

____

____

____

! ***Do not drink the water that you have treated!*** **At this stage of treatment, a municipal water treatment plant would add chlorine gas to kill any disease-causing organisms remaining in the water. Since our treatment process does not include chlorine, the water that you treated is not safe to drink.**

GUIDE TO ACTIVITY 21

Taking the Swamp Out of "Swamp Water"

◆What is happening?

This activity illustrates most of the major steps involved in purifying water for human consumption. While actual water treatment plants work with huge volumes of water daily, the processes that they use are fundamentally similar to those that you just completed.

Of the four processes illustrated, coagulation is the least familiar to most people. Alum (potassium aluminum sulfate) is used in many water treatment plants to help settle out small particles that are suspended in the water. Alum causes the very fine particles of soil to stick together (coagulate), forming floc. The largest clumps of floc quickly settle to the bottom of the water treatment tank. The floc that remains suspended in the water is easier to filter from the water than uncoagulated soil particles.

◆Time management

Depending on the ages of your students and the number and sizes of classes you teach, you may wish to set up the filtration part of the activity as a demonstration. The sand-pebble filters for this activity can be set up in advance. The same sand-pebble filter can be used to filter a number of different swamp water samples for successive classes if you rinse it with tap water between each use. If you plan to use the same sand-pebble filter several times, give it some additional support by clamping or taping it to a ring stand. This may help you to avoid a messy spill.

If the activity is done completely by the students (some of it is not done as a demonstration) allow at least two full class periods to perform and clean up—maybe a little more.

After the alum has been added, the water can be allowed to settle overnight if necessary. In general, the longer the floc settles, the clearer the water will be.

◆Preparation

Use sharp scissors to cut a cylinder from each of two of the soft drink bottles as follows:

• Cut the top off one of the bottles.

• Cut the base off the other bottle.

! Never push the point of the scissors toward your hand! Position your hand so that a slip of the scissors will not cause an injury!

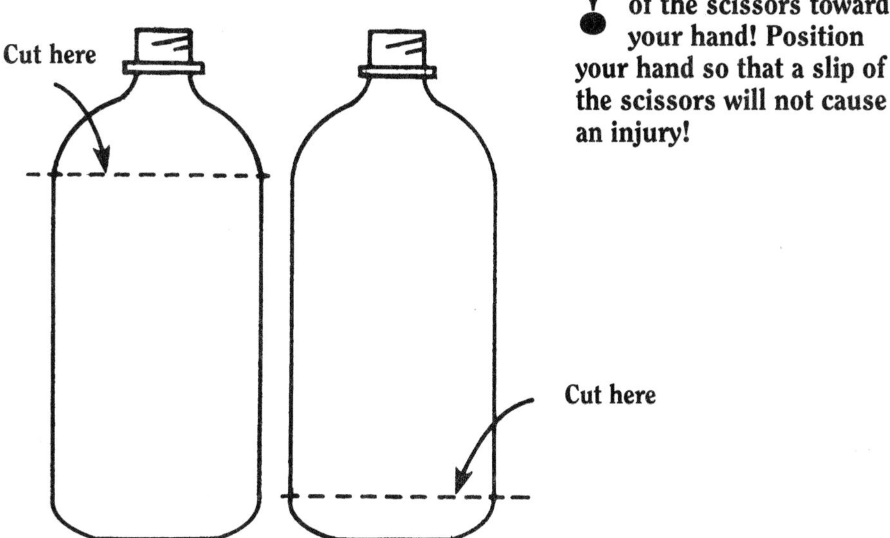

Assembling the sand-pebble filter in the cut-off bottle: Pebbles used for assembling the filter should be between 5 and 15 mm in diameter.

Either metal or nylon window screen may be used to hold

! If you think it is even remotely possible that students will drink the swamp water either before or after treatment, it would be best to prepare the swamp water by first baking the sediment in an oven at a high temperature (230° C, or about 450° F) for one hour to kill any pathogenic organisms present. Add the baked sediment to tap water.

the sand and pebbles inside the bottle. Such screen (available in hardware stores) can be easily cut with scissors. Be sure to prepare the 5 cm squares of screen ahead of time.

Rinse the sand-pebble filter with large volumes of clean water before using it to filter the swamp water. You can stop rinsing the filter when the clean tapwater running through the filter looks clear as it drips out of the filter.

Preparing swamp water:
Your swamp water supply may consist of water from a pond or creek (the muddier, the better), used aquarium water with extra dirt added, or "custom mixed" swamp water made by mixing a handful of mud or dirt into each liter of water.

About alum:
Alum is available in the spice sections of most supermarkets and also in drugstores. If the alum that you are using has large crystals, place it in a cup with a small amount of water and grind the crystals to a smaller size before adding the alum to the swamp water. Grinding the crystals into smaller particles increases the surface area available for sticking the suspended particles together. Iron salts such as ferric sulfide, ferrous sulfate, and ferric chloride may be used in place of alum.

• Set aside several liters of untreated swamp water as a control. Encourage students to compare the treated water that they produce with the untreated water.

• Any available large containers may be used to aerate the water and serve as a mixing container for the alum and swamp water. We recommend the use of cut-off plastic soft drink bottles because they are a convenient size, unbreakable, and free.

• A collection bucket should be provided for any untreated swamp water so plumbing does not get clogged.

◆Suggestions for further study

Plan a field trip to a local water treatment plant. Find out how (or whether) the plant removes: bacteria, lead or other heavy metals, nitrates, sulfides, calcium, other materials from the water.

Contact the state or local agency that tests water for contaminants. Have them test samples of tap water and the swamp water that you treated.

Add garlic powder to the swamp water and filter it out using some deodorizing charcoal and filter paper (coffee filters).

◆Answers

1. Answers will vary, depending on the water source used. Water from some sources may be smelly and/or muddy.

2. If the original water sample was smelly, the water should have less odor after aeration. Pouring the water back and forth allows some of the foul-smelling gases trapped in it to escape into the air of the room. You may see small bubbles suspended in the water and attached to the sides of the cylinders.

4. The rate of sedimentation depends on the water being used and the size of the alum crystals added. Large particles will settle almost as soon as stirring stops. Even if the water contains very fine clay particles, visible clumps of floc should form and begin to settle out by the end of the 20-minute period of observation.

6. After filtration, the treated swamp water should look much clearer than the untreated water does. It probably will not be as clear as tap water, but the decrease in the amount of material suspended in the water should be quite obvious. The treated sample should have very little odor when compared to the starting supply of swamp water.

MODULE 5

Investigating the Physical Properties of Water

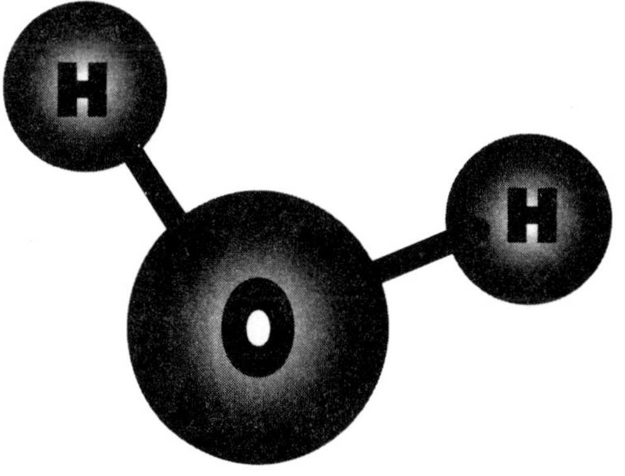

◆Introduction

Water is a "V"-shaped molecule composed of two hydrogen atoms bonded to one oxygen atom. Despite its chemical simplicity, water exhibits many unique physical properties:

• The range of temperatures routinely encountered on Earth allows water to exist in three different forms: as a solid (ice), a liquid, and a gas (water vapor).

• When air temperatures are below 0° C, the top layers of lakes and rivers freeze. Unlike most substances, water expands and becomes less dense when it forms a solid; therefore, ice floats, forming an insulating layer on top of the liquid water below. If ice sank, the newly exposed layer of water at the top would freeze, and it too would sink. Eventually many of our lakes and rivers would freeze solid, causing major changes in Earth's climate that would result in mass extinctions of many forms of plant and animal life.

• Water is often called "the universal solvent" because it can dissolve large amounts of many different materials and carry these materials for long distances. Many caves are formed as a result of water dissolving away underground rock. Because water can dissolve solids, the oceans contain not only salt, but enormous quantities of other materials as well.

• Capillary action allows water to "climb" to considerable heights against the force of gravity. Capillary action is the result of water's strong **cohesion** (the tendency of water to "hold itself together") and **adhesion** (the tendency of water to "stick to" different substances).

You will begin this workshop by watching three short segments of the TVOntario *Eureka!* science video series. The three programs deal with the general molecular properties of solids, liquids, and gases, and show how changing the temperature of a substance can cause it to change from one form to another.

You will also have the opportunity to participate in a number of brief hands-on investigations and demonstrations of the physical properties of water. These physical properties explain how water is recycled on Earth, and why water can and does reshape Earth's surface.

Your workshop leader may perform some of these activities as

Vocabulary

• **Adhesion:** The attraction between molecules that causes matter to cling or stick to other matter (as adhesive tape).

• **Cohesion:** The ability of a substance to stick to itself and "pull itself together."

demonstrations, or set up a series of lab stations for the activities. Individuals or groups may be asked to perform and present selected activities to the rest of the group. Whatever approach is taken, after observing and/or performing each demonstration, please answer the questions included in each activity.

SUMMARY OF MODULE 5

Investigating the Physical Properties of Water

◆Instructional objectives:

After completing the activities and readings for Module 5, you should be able to:

- describe what happens to individual molecules of water as water changes from a solid to liquid to a gas [Activities 22, 23, and 24]
- explain how factors such as temperature, wind, and humidity affect the rate at which water changes state [Activities 25, 26, and 27]
- demonstrate how capillary action allows water to move upward against gravity [Activity 28]
- demonstrate the cohesiveness of water molecules[Activities 28 and 29]
- explain how water's ability to dissolve materials makes the oceans salty and produces features geological features such as caves

◆Preparation

Study the following readings for Module 5:

Reading 17: What Is Water?

Reading 18: What Makes the Ocean Salty?

◆Activities

This module includes the following activities:

Activity 22: Eureka! #16: Molecules in Solids

Activity 23: Eureka! #17: Molecules in Liquids

Activity 24: Eureka! #18: Evaporation and Condensation

Activity 25: Changing States Is a Breeze

Activity 26: Feeling the Heat Makes My Molecules Dance

Activity 27: Saved by the Greenhouse

Activity 28: How Does Water Climb Sand Dunes?

Activity 29: Desert Rescue

ACTIVITY 22: VIDEOTAPE

EUREKA! #16: Molecules in Solids

◆Background on the *Eureka!* video series

The *Eureka!* videotapes you are about to see are part of a series of 30 animated science videos produced by TVOntario of Canada. Each segment is five minutes long and deals with a single science topic. The tapes are an excellent means of introducing a topic to your classes; they can also be used to review and summarize concepts that students are already familiar with.

The Eureka! style of presentation is both eye-catching and humorous. In addition, the scientific accuracy of the concept presentations is excellent. Both teachers and students find them fun and informative. The Recommended Audiovisual Materials section of this manual gives information on where to order your own copy of *Eureka!*

The following is a transcription of the summary of concepts presented in Program 16, "Molecules in Solids."

◆Concept summary

"All solids consist of little lumps of matter which are continuously vibrating to and fro in a lattice-work pattern."

"It is this latticework of little lumps that keeps the solid from falling apart."

"The scientific word for 'little lump' is molecule."*

◆Time management

Plan to spend at least double the running time of the film on this activity—this will allow time for an introduction and follow-up discussion.

◆Comments on the film

This *Eureka!* segment introduces the concept that all matter is composed of "little lumps" of material (atoms and molecules) that are in constant motion (dancing). In a solid, the molecules fit into a springy latticework where they alternately attract and repel each other. The spring symbolizes a complicated interaction between the electrons and the nuclei of the atoms and molecules. The film simplifies these interactions by showing pairs of molecules attached to one another by a spring. In reality, each molecule interacts with many other molecules.

Unlike the character in the film, students cannot shrink and look at individual moving molecules. For this reason, the concept of molecular motion is difficult for students to grasp, particularly in the case of solid materials. However, unless they accept the fact that all molecules are constantly in motion, students find that certain phenomena (such as changes in temperature, the ability of a liquid to flow, or the ability of a gas to disperse itself throughout a container) are very difficult to explain.

The concepts of molecular motion presented in the film apply to all substances. However, understanding molecular motion is particularly important when studying the physical properties of water. Water, unlike

---

most substances, readily changes state from a solid to a liquid to a gas at temperatures on Earth's surface.

The first scene of the film shows a man sniffing a bowl of soup, which is another way of saying that he is inhaling water vapor (among other things) escaping from liquid water (in the broth). What changes of state does the soup undergo during its preparation? Assume that the soup is made from frozen turkey broth. In order to do this, the cook heats the solid water (ice) of the broth, speeding up the motion of the water molecules. The ice in the broth melts, forming a liquid (warm soup). If the cook forgets and leaves the soup heating on the stove, its liquid water will evaporate or boil away, forming a gas (water vapor) that disperses throughout the air of the kitchen. Each step involves a change of state.

ACTIVITY 23: VIDEOTAPE

EUREKA! #17: Molecules in Liquids

◆Background

How are liquids different from solids? Ice is composed of the same types of molecules as liquid water, but a block of ice keeps its shape until it melts, while an identical quantity of liquid water flows down and takes the shape of the container it is poured into.

The following is a transcription of the summary of concepts presented in Program 17: "Molecules in Liquids."

◆Concept summary

"As the molecules in a solid get hotter, they vibrate faster and faster until their mutual force of attraction is no longer strong enough to hold them together."

"This causes them to slip out of their latticework pattern, which therefore falls apart."

"When the latticework of molecules in a solid has collapsed, we say that the solid has melted. It has changed from a solid state into a liquid state."*

◆Time management

Plan to spend at least double the running time of the film on this activity—this will allow time for an introduction and follow-up discussion.

◆Comments on the film

The differences in the properties of solids and liquids can be explained in terms of the molecular motion that occurs within them. The molecules composing a solid vibrate back and forth, but are held in place in a latticework.

The higher the temperature of a solid, the faster its molecules vibrate. For each solid, there is a specific temperature (that substance's melting point) at which the motions inside the solid become violent enough to break apart the latticework, and the molecules are then free to move. When the molecules escape the constraints of the latticework, the solid changes into a liquid or a gas.

A chocolate rabbit is used to demonstrate the process of changing from a solid to a liquid in this segment of *Eureka!*** Can you think of another solid (other than frozen water) that could so readily be transformed from a solid to liquid? Possibly not; most solids (like an iron pot that you might use to warm soup) do not melt readily, even if they are placed directly over a hot flame. Although iron can be melted (or even turned into vapor), doing so requires very high temperatures.

---

*Eureka! Produced by TVOntario© 1981

****Note:** Chocolate is *not*, in strict scientific terms, a solid. Even though it looks and feels solid, chocolate is generally classified as a *gel.* Gels have some properties like those of solids, and some properties that are more like those of liquids. These differences are not important in this particular example. The chocolate bunny was probably used because kids are familiar with (and adore) them. Who would care if a bunny made out of plain frozen water melted? Also, as this commentary points out, very few solid substances other than chocolate or ice would melt just from sitting out in the sun.

The process by which water changes from a solid to a liquid is not unique. However, the temperature at which this change of state occurs *is* unusual. Very few solids melt at similar temperatures. Most people would be willing to put their fingers into a container of melting ice, but would you volunteer to put your fingers into a container of melting steel or melting glass or melting rock?

ACTIVITY 24: VIDEOTAPE

EUREKA! #18: Evaporation and Condensation

Vocabulary

• **Cohesion:** The ability of a substance to stick to itself and "pull itself together."

• **Condensation:** Changing a gas into a liquid; for example, turning invisible water vapor into liquid water.

• **Evaporation:** Changing a liquid to a gas; for example, changing liquid water into water vapor.

• **Sublimation:** Formation of a gas from a solid, or vice-versa, without passing through the liquid phase.

◆Background

Evaporation and condensation are major steps in the water cycle. Without these processes, rain, snow, and other forms of precipitation would not occur. This segment of *Eureka!* examines evaporation and condensation at the molecular level.

The following is a transcription of the summary of concepts presented in Program 18: "Evaporation and Condensation."

◆Concept summary

"When molecules escape from a liquid, they spread out in all directions to form a gas or vapor."

"This process is called **evaporation:** the change of state from a liquid to a gas."

"When gas molecules are cooled, they go slower and crowd together more densely to form a liquid."

"This process is called **condensation:** the change of state from a gas to a liquid."*

◆Time management

Plan to spend at least double the running time of the film on this activity—this will allow time for an introduction and follow-up discussion.

◆Comments on the film

Cohesion is the force of attraction between the same types of molecules. Cohesion holds the molecules in a liquid together. In order for a liquid to evaporate, its molecules must overcome the force of cohesion and "fly" away from its surface. Molecules moving at high speed are more likely to overcome cohesion than are slowly moving molecules. The hotter a substance, the faster its molecules are moving. Therefore, the hotter the liquid, the more quickly it can change from a liquid to a gas.

At low temperatures, gas molecules condense—change from a gas to a liquid. Condensation is evaporation in reverse. Cooling the molecules of a gas slows their motions. Below a certain temperature, gas molecules will be moving slowly enough to be held together by cohesion. Then they begin to "stick" to other molecules of the same type, forming droplets of liquid.

If you cool a gas even more—below the temperature at which it condenses—the substance will change from a gas to a liquid to a solid. This is easy to do with water; by cooling it slightly below 0° C, you produce ice. Can you think of another common liquid that you can solidify so readily? Be careful—many of these liquids may, in fact, be mostly water. Liquids such as oil or pure alcohol do not freeze unless they are cooled to temperatures much lower than 0°C. Water is the best liquid for demonstrating changes of state.

Every day we encounter many examples of water changing states. We

---

**Eureka!* Produced by TVOntario© 1981

routinely change liquid water to a gas (steam) in teapots and coffee makers. On a cold day, you can see your breath. Cool air cannot hold much water vapor; the small cloud that forms in front of your face is the water vapor in your breath condensing as it cools from body temperature (about 37° C) to air temperature (usually around 5° C or lower when this happens).

Dew forms during the night on surfaces that become cool enough to condense the water vapor in the air. If the temperature is below freezing, the condensing water forms frost. (**Sublimation** is the term for when a gas changes directly to a solid—in this case water vapor to ice—or the reverse.) Dew and frost usually form on plants, but not on concrete; plants cool rapidly, but concrete stays too hot for condensation to occur.

A similar process occurs inside freezers and refrigerators. Especially during humid weather, frost (small ice crystals) tends to build up on the cooling coils. When you open a freezer door, humid air from the room fills the compartment. The water vapor in the air condenses on the cooling coils and then freezes, forming ice. Even "frost-free" freezers produce some frost; however, they avoid frost build up by heating the air inside the freezer once a day, thereby melting the frost. The liquid water that forms then runs out of the freezer compartment into an open pan, where it evaporates into the air in the room.

ACTIVITY 25 WORKSHEET

Changing States is a Breeze

◆Background

Water readily changes from a solid to a liquid to a gas. Each of these changes is called a *change of state.* What conditions make it easier for water to undergo a change of state?

◆Objective

To illustrate how wind affects the rate at which water changes state.

Materials

Each group will need

- an area of blackboard
- an electric fan
- 2 sponges
- a container of water

◆Procedure

1. Place the fan near the blackboard so that it blows only on one area of the blackboard; turn the fan on.

2. Moisten both sponges and squeeze out the excess water; both should be equally damp. No water should drip from the sponges.

3. Hold one sponge in front of the "breezy" location (an area of blackboard that the fan is blowing on); hold the second sponge in front of an area the fan is not blowing on.

4. Press the sponges onto the board at the same time so that each sponge leaves a wet "footprint" on the blackboard. Very little water should run down from the wet spots.

5. Your prediction: Which wet spot will disappear first: the water in the "breezy" location or the one where the fan is not blowing? Explain your prediction in terms of the motions of the water molecules "escaping" from the blackboard.

____

____

____

____

6. Which wet spot actually disappeared first? How accurate was your prediction?

____

____

____

Vocabulary

- **Adhesion:** The attraction between molecules that causes matter to cling or stick to other matter (as adhesive tape).
- **Cohesion:** The ability of a substance to stick to itself and "pull itself together."

GUIDE TO ACTIVITY 25

Changing States is a Breeze

◆What is happening?

When water evaporates, it changes from the liquid state to the gaseous state. The water molecules on the blackboard are "bouncing" up and down. When a molecule bounces far enough away from the board, it can "escape" from the wet spot as water vapor. The moving air from the fan strikes the bouncing water molecules and gives them a "push," causing them to move away from the board (evaporate) faster.

◆Time management

Allow 20 to 30 minutes to perform this activity and discuss it.

◆Preparation

You may find it easier to do this activity as a demonstration, rather than having groups of students do the activity. If the lighting in the room is at the proper angle, the class should be able to see the spots disappear without leaving their seats.

• Substitute materials: A hair drier set on the "fan only" setting is a good substitute for an electric fan. You could also use a piece of cardboard to fan one of the spots by hand. If you do not have a blackboard, or if the chalkboard that is available will be harmed by water, press the sponges on a piece of smooth masonite.

◆Suggestions for further study

Ask students to measure the amount of time it takes for each wet spot to disappear. Why do clothes dry more quickly outdoors if there is a breeze? Talk about the energy conservation implications of nature's great "outdoor dryer."

◆Answers

5. Predictions will vary, but most students will probably predict that the water in the "breezy" location will evaporate first.

Air consists of molecules of different kinds of gases—mainly nitrogen and oxygen. The gas molecules of the air being moved by the fan bump into the water molecules, giving them a "push." This helps them to overcome the **cohesion** that holds them to the other water molecules and the **adhesion** that holds them to the board.

As the water molecules "escape" from the blackboard, the motion of the air carries the escaping water vapor away. This makes it less likely that any escaping molecule of water will bump back onto the board and "reattach" itself to the spot of water on the board. It also makes it easier for other water molecules to escape from the blackboard, since the air becomes "drier" as the water vapor is carried away. For these reasons, the spot in the "breezy" location disappears first.

Sample data

The water in the breezy location disappeared in *one* minute.

The water in the calm location disappeared in *four* minutes.

ACTIVITY 26 WORKSHEET

Feeling the Heat Makes My Molecules Dance

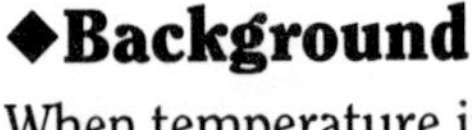

◆Background

When temperature increases, the molecules in a substance move faster. Will increasing the temperature make it easier for water to undergo a change of state?

◆Objective

To explain how temperature affects the rate at which water changes state.

Materials

Each group will need

- an area of blackboard
- an electric lamp with a 100 watt bulb
- 2 sponges
- a container of water

Vocabulary

- **Adhesion:** The attraction between molecules that causes matter to cling or stick to other matter (as adhesive tape).
- **Cohesion:** The ability of a substance to stick to itself and "pull itself together."

Keep the electric lamp away from water!

◆Procedure

1. Place the lamp as close as possible to the blackboard and turn it on. Leave the bulb burning until an area of the surface of the blackboard about 20 cm in diameter is warm (*not* hot) to the touch. (This may require three to five minutes—check the blackboard after a few minutes and see if it feels warm.)

2. When the board is warm, turn off the lamp and move it away from the blackboard. Moisten the sponges and squeeze out any excess water. Both sponges should be equally damp. No water should drip from them.

3. At the same time, press one sponge against the warm spot on the blackboard and the second sponge against an area that was not warmed by the lamp.
Both sponges should leave damp "footprints" on the board, but very little water should run down from the damp spots.

4. Your prediction: Which wet spot will disappear first—the one on the heated area of the board or the one on the unheated area? Explain your prediction in terms of the motions of the water molecules escaping from the blackboard.

____

____

____

5. Which wet spot disappeared first? How accurate was your prediction?

____

____

____

GUIDE TO ACTIVITY 26

Feeling the Heat Makes My Molecules Dance

◆What is happening?

When water evaporates, it changes from a liquid to a gas. This change of state requires an input of energy. Water can obtain this energy from a warm surface or by absorption of electromagnetic energy (such as sunlight).

Warming the blackboard with the lamp provides a source of heat energy for the water placed on that area. This energy makes the molecules of water placed on the warm spot move more rapidly than the water molecules placed on the cool section of the blackboard. The warmer area of blackboard dries first.

◆Time management

Allow 20 to 30 minutes to perform this activity and discuss it.

◆Preparation

Clamp lamps work well for this activity.

Bulbs larger than 100 watts get extremely hot and are not recommended for use in this activity.

The blackboard only needs to be slightly warm to the touch before the water is placed on it. A hair drier set on "high" can be used to warm the area of blackboard instead of a light bulb. If you do not have a blackboard, or if the blackboard that is available will be harmed by heat or water, press the sponges on a piece of smooth masonite or use cardboard.

◆Suggestions for further study

Ask the students if a fan will cool or warm the area. Then ask them to predict what will happen based on their observations from the previous experiment. Now repeat the experiment while fanning the warm area with a piece of paper. Explain that the fan moves water-saturated air away from the board, which allows unsaturated air in near the board, thereby increasing the rate of evaporation. Evaporation absorbs heat which lowers the temperature.

◆Answers

4. Answers will vary, but most students will probably predict that the water in the heated location will evaporate first.

Water molecules are always in motion. The speed with which they move depends on the temperature of the water—the higher the temperature, the faster the molecular motion. The water placed on the heated area absorbs heat from the blackboard; its molecules begin moving faster than the water on the unheated area. The faster-moving molecules on the heated area are more likely to escape from the blackboard by overcoming the **cohesion** that holds them to the other water molecules and the **adhesion** that holds them to the board. Therefore, the water disappears from the heated area first.

Sample data

The water in the heated area disappeared in *one* minute.

The water in the unheated area disappeared in *four* minutes.

Note: These times will vary depending on the temperature of the heated area of blackboard compared to room temperature, the amount of water in the wet spots, and the relative humidity of the room. The evaporation times may range from 30 seconds to 3 minutes for the warm area, and from 1 to 5 minutes for the cool area.

ACTIVITY 27 WORKSHEET

Saved by the Greenhouse

◆Background

The amount of water in the air influences the rate of evaporation of water. The more water vapor already in the air, the slower will be the rate of evaporation.

Materials

Each group will need

- the top of a cardboard shirt (gift) box
- 4 paper towels (divided into two long sheets of 2 towels)
- scissors
- plastic food wrap
- masking tape
- a stapler
- a metric ruler
- 2 sponges
- an area of blackboard
- a container of water

Vocabulary

- **Transpiration:** The process by which living plants give off water vapor into the atmosphere.

◆Procedure

1. Make a greenhouse as follows:

Cut a window approximately 10 cm long by 15 cm wide in the box top. Cover the window with clear plastic wrap, and tape it down tightly on all sides. Many gift boxes are thin and weak; you may have to strengthen the corners of the box with tape.

Roll the paper towels into long cylinders. Staple them on the inside of the box, one on each side.

Dampen the paper towels by carefully rubbing them with a wet sponge.

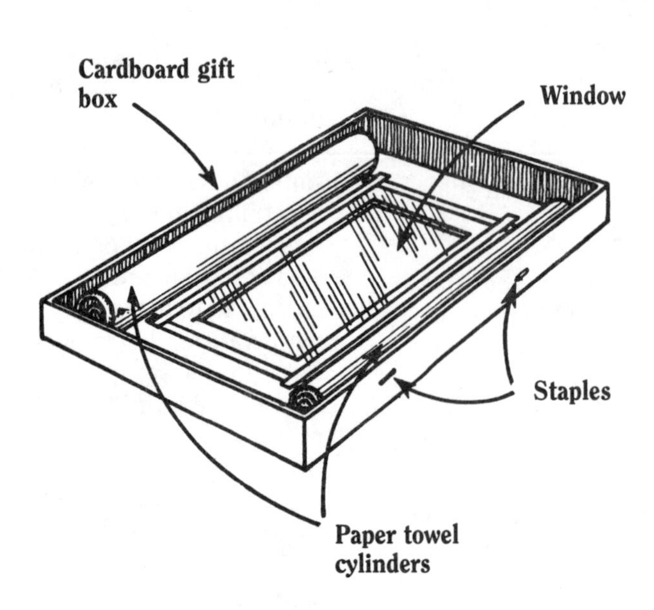

2. Moisten the sponges and squeeze out the excess water. They should be thoroughly wet, but not dripping.

3. At the same time, press the sponges against two separate areas of blackboard. Each sponge should leave a wet "footprint" on the board, but very little water should run down from the wet spot.

4. Quickly place the greenhouse (with paper towels facing the board) over one of the wet spots. Tape the greenhouse securely to the blackboard, using tape to seal any openings between the greenhouse and the board.

5. Your prediction: Which wet spot do you expect to disappear first—the one covered by the greenhouse or the unprotected one? Why?

6. Which wet spot actually disappeared first—the one covered by the greenhouse or the unprotected one? Was your prediction correct?

7. Have you ever walked through a large greenhouse where plants were being grown? Did the air seem damp inside? Why do you think that this is the case?

8. Be sure to clean any remains of tape from the blackboard when you are finished with this activity.

GUIDE TO ACTIVITY 27

Saved by the Greenhouse

◆What is happening?

The air inside the greenhouse quickly fills with water vapor (100% of relative humidity) from the damp paper towels. Since the air inside is almost saturated with water vapor, the water molecules placed on the blackboard have a harder time escaping from the board (changing from the liquid state to the gaseous state). The spot under the greenhouse evaporates more slowly than the spot that is exposed to the air in the room.

Commercial plant growers take advantage of this effect. The plants constantly take up water from the soil, and give off (transpire) water through tiny pores on their leaves. Greenhouses hold this water vapor inside, helping to keep the plant moist without constant watering.

◆Time management

The time necessary for this activity will vary depending on how much time it takes the students to construct the greenhouse and the amount of time it takes the water to evaporate. Plan to use between one and a half to two class periods.

If you perform "Saved by the Greenhouse" immediately after the activity "Feeling the Heat Makes My Molecules Dance," be sure to move away from the warm area of blackboard—placing the water spots on a warm area may produce confusing results.

◆Preparation

Avoid placing one spot in a sunlit area of the board and the other in a shaded area. The sunlit area may be much warmer than the shady area.

- If you do not have a blackboard, or if the chalkboard that is available will be harmed by water, press the sponges on a piece of smooth masonite or cardboard.
- Be sure to have students remove all pieces of tape from the blackboard after completing the activity.

◆Suggestions for further study

Cover a large plant with a clear plastic bag, sealing the bag tightly around the base of the plant's stem. Be sure the plant has been watered (however also make sure that there is not any water standing in the pot) and allow to stand for a day. Repeat with different sizes and species of plants.
Each student can make their own mini-greenhouse by cutting a 2 L clear plastic soda bottle apart about 10 centimeters from the base; filling the base with a mixture of potting soil, sand, and charcoal; and adding a small plant. Sprinkle the plant and soil with water, but do not soak, and replace the upper portion of the bottle, sealing with heavy clear tape. Keep the cap on the bottle and watch the plant grow and recycle the water.

◆Answers

5. Predictions will vary, but most students will probably predict that the water that is not covered by the greenhouse will evaporate first.

The total amount of water vapor that a given volume of air can hold depends on the temperature of the air. The higher the temperature, the more water vapor air can hold. The air inside the greenhouse is almost saturated with water vapor from the wet paper towels. In other words, the air inside already contains about as much water vapor as it can hold. This means that there is not much space inside the greenhouse for the water molecules from the blackboard to occupy. Therefore, the liquid water inside the greenhouse evaporates from the blackboard more slowly than the water that is exposed to the air.

6. Most people will observe that the spot inside the greenhouse evaporates more slowly than the unprotected spot.

7. Commercial greenhouses feel damp inside because the plants growing there release large amounts of water vapor into the air of the greenhouse through the process of **transpiration.** Since the water vapor cannot escape from the greenhouse, the air inside becomes saturated with water vapor. Once it is saturated with water vapor, the air in the greenhouse cannot readily absorb additional moisture, so the rate of water loss from the plants slows down. This decreases the amount of water that the plants need to continue healthy growth.

Sample data

The liquid water in the uncovered area disappeared in *4* minutes.

The liquid water in the "greenhouse" disappeared in *12* minutes.

Note: These times will vary, depending on how tightly the greenhouse is taped to the blackboard, the temperature, the amount of water vapor already in the room, and the amount of water on the towels and in the wet spots themselves. In a very humid room, the uncovered spot may take almost as long as the spot inside the greenhouse to disappear.

ACTIVITY 28 WORKSHEET

How Does Water Climb Sand Dunes?

Materials

Each group will need

- a flexible, transparent (or translucent) plastic cup
- scissors
- 200 ml of water
- fine-grained dry sand—enough to fill the cup
- a pie pan or dish with a rim at least 1 cm high
- metric ruler

Vocabulary

- **Adhesion:** The attraction between molecules that causes matter to cling or stick to other matter (as adhesive tape).
- **Cohesion:** The ability of a substance to stick to itself and "pull itself together."

◆Background

Can water really defy gravity? Both **adhesion** and **cohesion** are at work when water rises through dry sand. This upward movement of water is called **capillarity** or **capillary migration.**

◆Objective

To demonstrate how capillary action allows water to move upward against gravity.

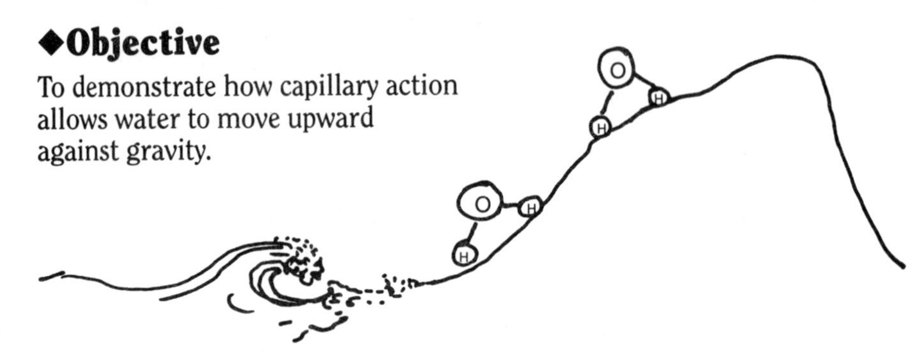

◆Procedure

1. Use the scissors to cut the bottom out of the cup, producing a cylinder that is open at both ends. Check the inside to be sure that it is dry. Wipe off any water droplets that are present.

2. Place the cylinder you made in the pie pan. Fill the cylinder completely with dry sand. You now have a dry column of sand sitting inside the cylinder in the pie pan.

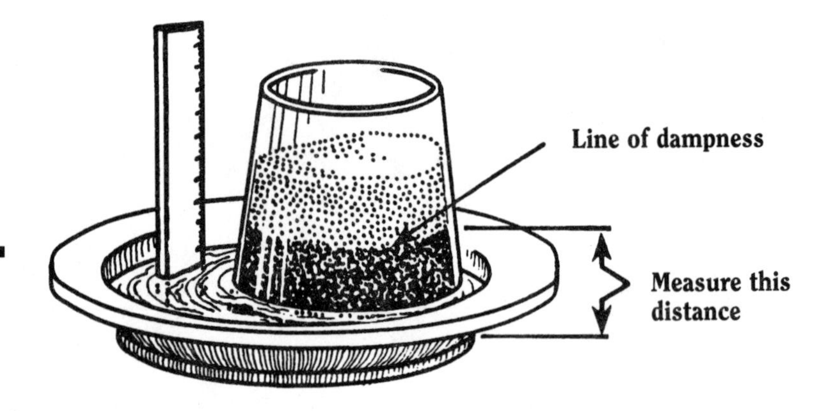

3. Gently pour the water directly into the bottom of the pie pan to a depth of 1 cm. (Use a metric ruler to measure the depth.)

Do *not* move the cylinder, pour water directly on the sand, or disturb the sand.

4. Observe how many centimeters the water rises through the sand after: 1, 2, 5, 10, and 30 minutes.

Measure the distance from the bottom of the cylinder to the top of the line of dampness in the sand.

Record your observations in the data table at left.

Height of water in the "sand dune"

Time elapsed after water was added	Height in cm
1 minute	______cm
2 minutes	______cm
5 minutes	______cm
10 minutes	______cm
30 minutes	______cm

5. Describe what happens when the water is poured at the base of the cylinder containing the sand dune.

____

____

____

6. After 30 minutes, carefully pick up the cylinder while holding it over the plate. Does the sand pour out as individual grains or in a lump? Does this observation give evidence that the forces of cohesion and adhesion are acting on the sand and water?

____

____

____

GUIDE TO ACTIVITY 28

How Does Water Climb Sand Dunes?

Vocabulary

• **Capillarity:** The process by which water rises through rock, sediment, or soil. The cohesion between water molecules and adhesion between water and other materials "pull" the water upward; also called capillary action.

◆What is happening?

Both **adhesion** and **cohesion** are at work when water rises through dry sand. This upward movement of water is called **capillarity** or **capillary migration**. Adhesion between the molecules of water and the individual particles of sand allows water to coat each sand particle that it touches. The force of cohesion holds the water molecules together, so that as water coats each sand particle and rises up the "sand dune," more water is pulled up from below. Once the sand is completely wet, the combined forces of adhesion and cohesion are strong enough to bind it into a lump that can be removed from the cup in one piece.

◆Time management

Allow one class period for this activity, but plan to do something else while waiting to collect data. You could assign a reading from Module 5 or perform the next activity, "Desert Rescue," while taking the readings of the water rising through the column of sand.

◆Preparation

Water does not rise through the sand perfectly evenly; the visible line of moisture will be higher in some places and lower in others. Encourage your students to use a consistent technique when measuring the height of the water. They can measure the height of the water at its highest level, lowest level, or at an "average" level above the pan. The important point is that they should use the same criterion for locating the water level every time they make a measurement.

If you think it will be at all difficult or dangerous for the students to cut the plastic cups; it may be better to cut them yourself before performing the activity.

You can make "cylinders" with vertical sidewalls by cutting off sections from 1 L plastic soft drink bottles. Some students find it easier to make accurate measurements using this type of cylinder.

Even "dry" sand may contain enough water to affect the results of this activity. Your students will get more consistent results if you reduce the water content of the sand by air-drying it on a sunny windowsill for several days before class, or by heating it in the following manner:

1. Spread the sand on a cookie sheet in a layer about 1 cm thick.
2. Place the cookie sheet in a hot (200° C, or about 400° F) oven for 20 minutes. Stir the sand occasionally.
3. Allow sand to cool to room temperature before using it.

◆Suggestions for further study

Capillary action is an important means of moving water through all types of sediments—not just through sand. Repeat this activity using silt, topsoil, fine sand, coarse sand, or pebbles. How far does the water rise? How does the particle size affect the results?

Allow the water to continue rising overnight in these sediments. Does the water eventually reach the top of the column?

Construct a tall (1 m or higher) sand column inside a piece of clear plastic tubing or plastic pipe. How high does water rise? How long does it take?

Discuss with your class possible reasons why the water does not rise evenly through the columns of soil. Encourage students to identify variables that may affect the way the water rises. When they have identified several variables, have them design and perform experiments to test their hypotheses on what controls the rate and uniformity of rising water.

◆Answers

Data: The rate at which water rises varies depending on the size of the sand particles, how closely they are packed, and the amount of water already present in the sand before water is added to the dish containing the column of sand. The smaller the grains of sand, the faster (and more obvious) will be the rise of the water.

5. The water rises up the sand column. You can clearly see a dividing line inside the cylinder between the wet sand below and the dry sand above. The wet sand extends several centimeters above the surface of the water.

6. The wet sand comes out as a lump. Adhesion allows the water coating the sand grains to hold onto the individual grains. Cohesion between the water molecules helps to hold the sand and water together, forming a lump of sand.

ACTIVITY 29 WORKSHEET

Desert Rescue

Materials

Each group will need

- a "canteen"—a beaker with a pouring spout (or a measuring cup)
- a metal or plastic cup
- 1.5 m of cotton twine
- approximately 200 ml of water
- a meter stick

◆Background

The story:
While hiking through the desert, you encounter a prospector trapped in a vertical ventilation shaft of an abandoned mine. He has injured both arms during his fall, so he cannot climb out, and his head is about 1.5 m below ground level. You cannot reach him and have nothing to help you pull him out. He needs water to survive, but you must keep some of your water for your own use while you hike to the nearest road to get help.

The problem:
You have only a metal cup, a small canteen of water, and a ball of very weak twine to help you rescue the prospector. The twine is too weak to allow you to lower and raise the canteen, and the prospector could not throw the canteen back up to you if you dropped it to him. Dropping the canteen is too dangerous anyway—it might hit him in the head, or the water could spill out and both of you might die of thirst.

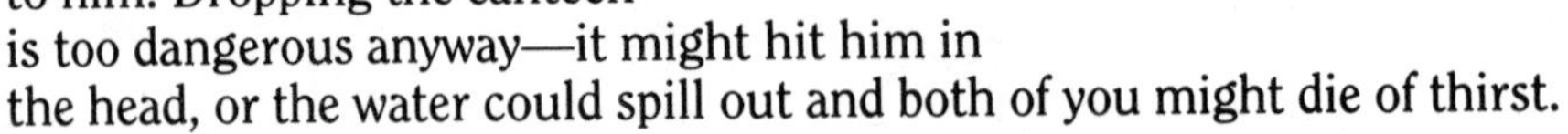

- Is there a way that you can use your knowledge of the physical properties of water to (1) keep the canteen with you, (2) get water to the prospector, and (3) avoid spilling any of your limited supply?

◆Objective

To demonstrate the cohesive nature of water molecules.

◆Procedure

1. Apply your knowledge of the physical properties of water to the problem stated above. Discover a way that you can transfer about half of the 200 ml of water from the "canteen" to the prospector about 1.5 m below. Remember, both your lives depend on your ability to transfer the water without wasting a drop.

2. When you come up with a way to transfer the water, describe how you would get water to the prospector:

GUIDE TO ACTIVITY 29

Desert Rescue

◆What is happening?

Cohesion holds molecules of liquid water together. When you turn on a faucet, the water forms a continuous stream from the faucet to the sink (if the water is not moving too fast or too slowly). However, as water falls it accelerates. Falling water soon reaches a speed at which cohesion is no longer sufficient to hold it in a single stream. Instead, the water breaks up into droplets. Drops of water are spherical because of cohesion.

Water also exhibits the property of adhesion. It adheres to cotton twine very well. If you wet the cotton twine, and combine the force of cohesion with the force of adhesion between the water and the twine, you can easily pour the water out of the canteen to the prospector below. Using this system, the water will even travel quite a distance horizontally along the twine.

Unless you are willing to waste much of your supply, you cannot get water to the prospector by simply pouring it into the ventilation shaft and aiming for the metal cup. The cohesion between the water molecules is not strong enough to hold the water together in a single stream all the way to the prospector's cup, so the stream will break up into droplets. Some of the droplets will miss the cup. Much of the water will splash out even if it lands in the cup.

◆Time management

Let the students think about this problem for a while—perhaps overnight. Encourage them to discover a solution to the problem, rather than just giving them the solution.
Plan to use between one and two class periods for this activity.

◆Preparation

Before performing this activity, cut a 1.5 m length of twine for each group that will be participating.
Cotton twine that is fairly thick (about the diameter of a pencil lead) works best for this activity. Pre-wetting the twine may reduce the amount of dripping that occurs while the water is being poured.

◆Suggestions for further study

Show high speed stroboscopic pictures of liquids dropping. Use NASA film "Zero G" to show the properties of liquid cohesion. Ask why the water forms into a sphere (this is the shape which allows for the least amount of surface area). Why are raindrops tear-shaped? (air motion draws the drop out of its spherical shape).

Place drops of water on various surfaces (table top, wax paper, plastic wrap, paper towel, etc.) and discuss the observations of the shape of the drops in terms of adhesion and cohesion.

◆Answers

2. To get water to the prospector, do the following:

- Tie the twine to the cup with a bow and lower it to the prospector. Try to avoid hitting the prospector on the head.
- Have the prospector untie the twine from the cup and hold the twine in the bottom of the cup.

Place your end of the twine on top of the pouring spout of your canteen. Pull the twine taught, and slowly pour the water onto the twine. The stream of water will flow into the injured man's metal cup.

The water forms a continuous stream along the twine. Cohesion holds the water molecules together; adhesion between the twine and the water keeps the water attached to the twine.

Note: It is not essential to use the metal cup to transfer the water to the prospector. If his arms are severely injured, he can just hold the string between his teeth while you pour, and swallow the water as it flows into his mouth.

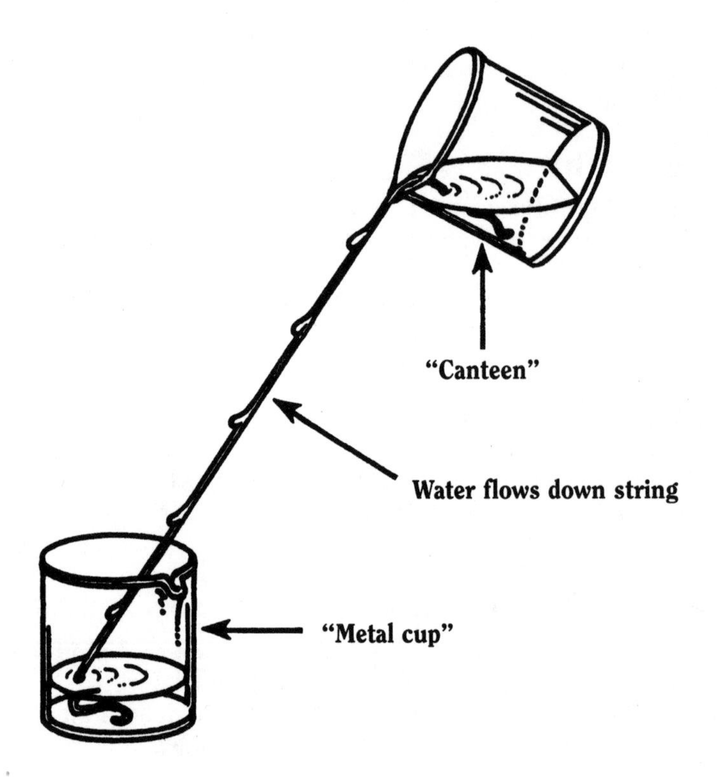

EARTH: THE WATER PLANET

Readings

◆Introduction

The following readings review and expand upon the concepts introduced in the activities of these five modules. The readings include essays written especially for these modules, reprinted newspaper accounts dealing with current water issues, and booklets produced by several different federal agencies.

Depending on the needs and interests of your students, you may wish to duplicate selections from these readings or obtain copies of the government booklets for student use. However, you should be alert to the fact that while the scientific content of these booklets is generally accurate, some of the older federal publications contain statements that depict men and women in sexually stereotyped roles or use "man" in a generic sense to mean "human beings."

READING 1

Groundwater: What, Where, How and Why

◆What is groundwater?

Water found anywhere under Earth's water table is called **groundwater**. It is simply the subsurface water that fills the pores or cracks in soils and rocks in the saturated zone beneath the water table.

Ancient chronicles show that people have long known that there is a lot of water underground. Only within recent decades, however, have scientists and engineers learned to estimate how much groundwater is stored underground, and begun to document its vast potential for use. We now know that in terms of storage at any one instant in time, groundwater is the largest single supply of fresh water available for use.

An estimated 4.2 cubic km (about one million cubic miles) of the world's groundwater is stored within 800 m (one-half mile) of the land surface. Only a fraction of this reservoir of groundwater, however, can be practicably tapped and made available through wells and springs. The total amount of groundwater is more than 30 times greater than the nearly 125,000 cubic km (30,000 cubic miles) volume in all the fresh-water lakes; it is more than 3000 times greater than the water flowing in all the world's streams at any given time.

◆Where is groundwater found?

Some water underlies the Earth's surface almost everywhere; there is water beneath hills, mountains, plains, and deserts. It is not always accessible, however, or fresh enough for use without treatment. Groundwater may seep to the land surface in a marsh, but in some arid areas it may lie many hundreds of feet below the surface and be difficult to locate and use.

Groundwater is stored in—and moves slowly through—moderately to highly permeable layers of rock or sediment called **aquifers.** The word aquifer comes from the two Latin words, *aqua*, or water, and *ferre,* to bear or carry. Aquifers literally carry water underground. After entering an aquifer, water moves slowly toward lower-lying places. Water is discharged from aquifers as springs, or it may seep from an aquifer into streams, or it can be removed by drilling a well into the aquifer.

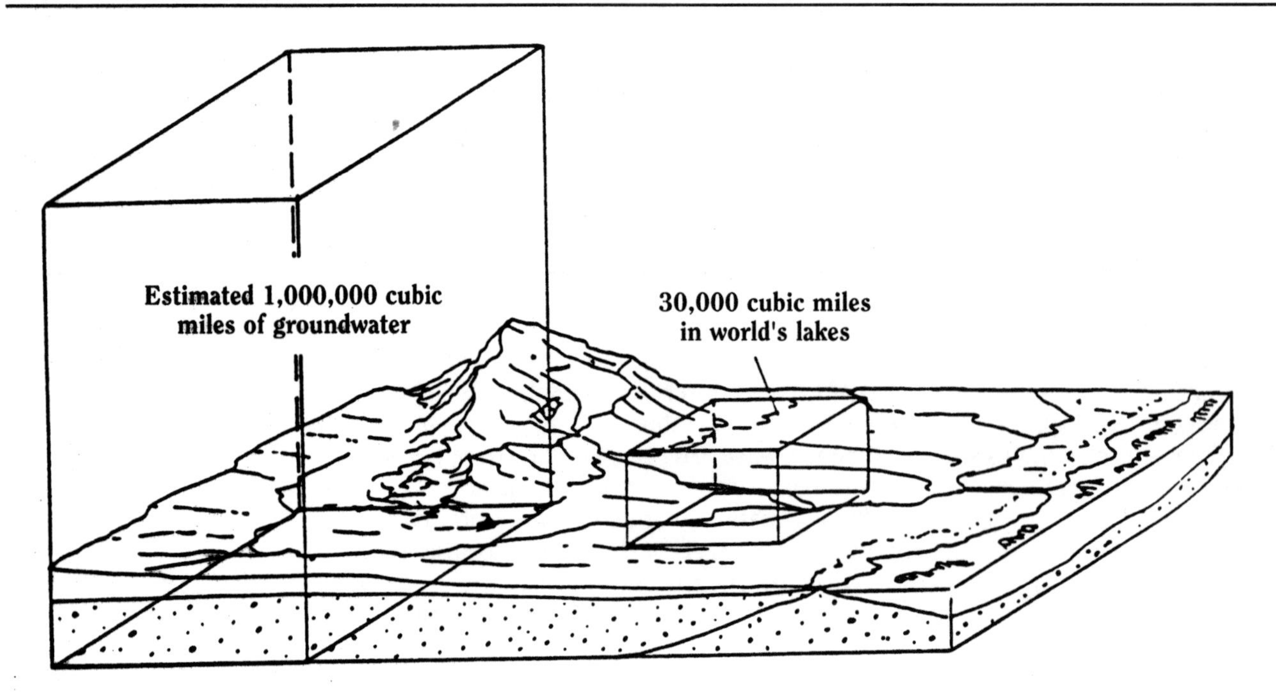

An aquifer may consist of a layer of sediment such as gravel or sand, a layer of sandstone or cavernous limestone, a rubbly top or base of lava flows, or even a large body of massive rock (such as fractured granite) that has sizable openings. An aquifer may be only a few feet thick or it may be tens or hundreds of feet thick. It may lie a few feet below the land surface or thousands of feet below the surface.

If water is to move through rock or sediment, the material's pores must be connected to one another. If the pore spaces are connected and are large enough that water can move freely through them, the material is said to be **permeable.** Rock or sediment that will yield large volumes of water to wells or springs must have many interconnected pore spaces or cracks. A rock with few pore spaces, such as granite, may be permeable if it contains enough sizable and interconnected cracks or fractures.

The quantity of water a given type of rock or sediment will hold depends on its **porosity**—a measure of the amount of pore space between the grains or cracks in the material that can fill with water. Generally, the porosity and permeability of rocks decrease as their depth below land surface increases; the pores and cracks in rocks at great depths are closed or greatly reduced in size because of the weight of overlying rocks.

Adapted from the U.S. Department of the Interior/Geological Survey booklet "Ground Water" (U.S. Government Printing Office: 1986-491-402/04).

READING 2

The Ins and Outs of Groundwater

There is an enormous amount of water underground. Unfortunately, the distribution of this water resource is irregular; the groundwater supply cannot always meet the demands placed upon it. Some areas of the United States already face serious regional water shortages because groundwater is being used faster than it is naturally replenished by water entering the ground; in other areas, groundwater is being polluted by human activities. The processes by which we draw water from the groundwater supply and the ways in which the supply is replenished are discussed in the following sections.

◆How groundwater reaches Earth's surface

When surface water is scarce or nonexistent, finding a groundwater source may be the difference between life and death. Long before the beginning of recorded history, human beings dug wells to insure an adequate water supply. Our ancestors may have learned this trick from other species such as horses and elephants that also dig holes to obtain water.

A well may be regarded as nothing more than an extra large pore in the soil or rock. A well dug or drilled into water-saturated rocks will fill with water approximately to the level of the water table. If water is pumped from a well, gravity will force water to move from the saturated rocks into the well to replace the pumped water.

Pumping may not be required to obtain water from **artesian wells.** In areas where the groundwater is trapped between layers of impermeable materials (such as clay or shale), the water may be confined under

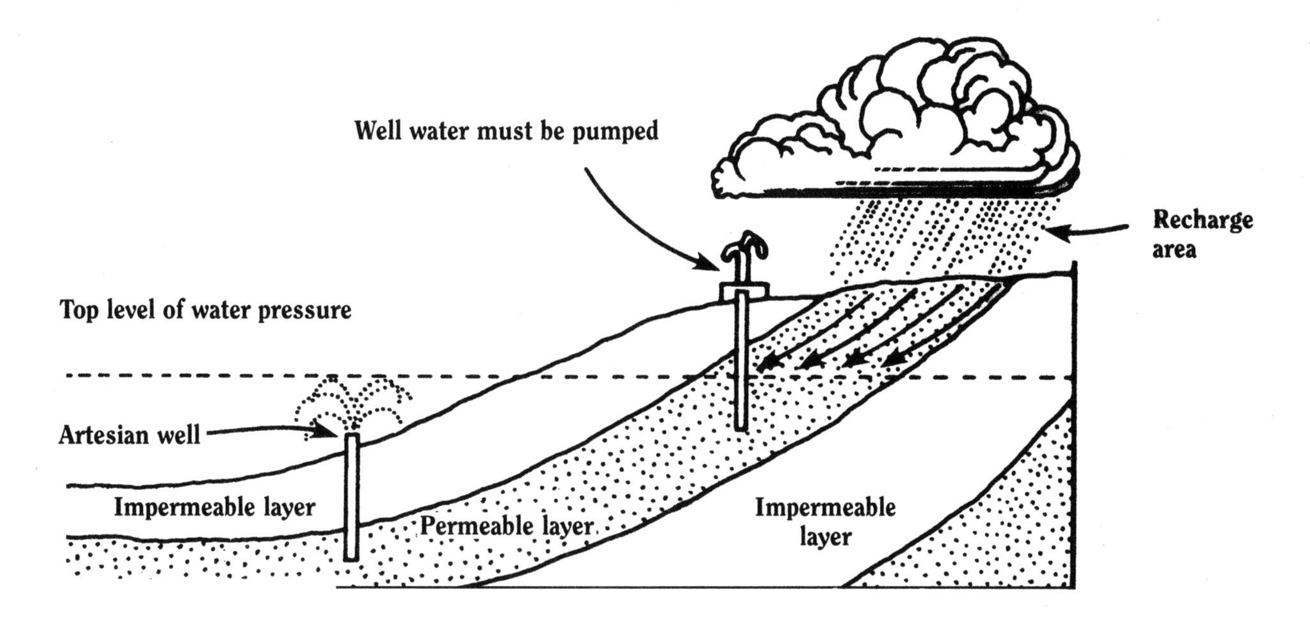

pressure. This type of confined aquifer is called an **artesian aquifer,** and water contained in the aquifer is said to be under artesian pressure. If such a confined aquifer is tapped by a well, the artesian pressure will push the water above the top of the aquifer. In some aquifers, the artesian pressure is great enough to cause the water to flow from the well onto the land surface.

A **spring** is the result of an aquifer being filled to the point that the water overflows onto the land surface. Springs are classified in several different ways: according to the geologic formation from which they obtain their water, such as limestone springs or lava-rock springs; according to the amount of water they discharge—large or small; according to the temperature of the water—hot, warm, or cold; or by the forces causing the spring—gravity or artesian flow.

Thermal springs are ordinary springs except that the water is warm and, in some places, hot. Many thermal springs occur in regions of recent volcanic activity and are fed by water heated by contact with hot rocks far below the surface. Such are the thermal springs in Yellowstone National Park. **Geysers** are thermal springs that erupt intermittently and to differing heights above the land surface. Eruptions occur when water deep in a thermal spring is heated enough to change into steam; this steam expands, pushing the liquid water above it into the air. Some geysers are spectacular and world famous, such as Old Faithful in Yellowstone National Park in Wyoming.

◆How does water enter the ground?

Groundwater is replenished by precipitation. When rain falls or snow melts, some of the water evaporates; some flows overland and collects in streams; and some infiltrates into the pores or cracks of the soil and rocks. The first water that enters the soil replaces water that has been evaporated or used by plants during a preceding dry period.

After the water requirements for plants and soil are satisfied, any excess water will percolate downward to the **water table**—the top of the zone below which all openings in and between rocks and soil particles are saturated with water. Below the water table, the water moves horizontally through the aquifer to streams, springs, or wells from which water is being withdrawn.

Natural refilling (or recharge) of aquifers is a slow process because groundwater moves slowly through the unsaturated zone and the aquifer. Water at very shallow depths may have been underground for just a few hours; at moderate depth, it may be 100 years old; and at great depths or after having flowed long distances from places of entry, water may be several thousands of years old.

The rate of recharge is an important consideration for groundwater users. It has been estimated, for example, that if the aquifer that underlies the High Plains of Texas and New Mexico—an area of slight precipitation—were emptied, it would take centuries to refill the aquifer at the present small rate of replenishment. In contrast, a shallow aquifer in an area of substantial precipitation may be replenished almost immediately by one rainstorm.

In some areas, attempts are being made to replenish aquifers artificially. For example, large volumes of groundwater used for air conditioning are returned to aquifers through recharge wells on Long Island, New York. Although some artificial-recharge projects have been successful, many have been disappointments; there is still much to be learned about different groundwater environments and their receptivity to artificial-recharge practices.

Adapted from the U.S. Department of the Interior/Geological Survey booklet "Ground Water" (U.S. Government Printing Office: 1986-491-402/04).

READING 3

Quality and Protection of Our Groundwater

For our nation as a whole, the chemical and biological character of our groundwater is acceptable for most uses. However, the quality of groundwater in some parts of the country, particularly shallow groundwater, is declining as a result of human activity.

Further development of energy, mineral, and agricultural resources is dependent largely upon adequate, inexpensive water supplies. Groundwater resources will become even more valuable in the years ahead as the nation copes with growing natural resource and environmental problems including increased water demands.

Groundwater is less susceptible to bacterial pollution than surface water because the soil and rocks through which groundwater flows screen out most of the bacteria. However, bacteria occasionally find their way into groundwater, sometimes in dangerously high concentrations. But freedom from bacterial pollution alone does not mean that the water is fit to drink. Many unseen dissolved mineral and organic constituents are present in groundwater in various concentrations. Most are harmless or even beneficial; others are harmful, and a few may be highly toxic.

Water is an excellent solvent; it dissolves minerals from the rocks with which it comes in contact. Groundwater may contain dissolved minerals and gases that give it the tangy taste enjoyed by many people. Without these minerals and gases, the water would taste "flat" like distilled water. On the other hand, water from some wells and springs contains very large concentrations of dissolved minerals and cannot be tolerated by most living organisms. Many parts of the nation are underlaid by highly saline groundwater that has only very limited uses.

Dissolved mineral constituents can be hazardous to animals or plants in large concentrations; for example, too much sodium in the water may be harmful to people who have heart trouble. Boron is a mineral that is good for plants in small amounts, but is toxic to some plants in only slightly larger concentrations.

Water that contains a lot of calcium and magnesium is said to be **hard.** The hardness of water is expressed in terms of the amount of calcium carbonate—the principal constituent of limestone—or equivalent minerals that would be formed if the water were evaporated. Very hard water is not desirable for many domestic uses; it will leave a scaly deposit on the inside of pipes, boilers, and tanks.

Hard water can be softened at a fairly reasonable cost, but it is not always desirable to remove all the minerals that make water hard. Extremely soft water is likely to corrode metals, although it is preferred for laundering, dishwashing, and bathing. In many places, groundwater contains excessive amounts of iron. Iron causes reddish stains on plumbing fixtures and clothing. Like hardness, excessive iron content can be reduced by treatment.

In recent years, the growth of industry, technology, and population has increased the stress upon both our land and water resources. In many places, the quality of groundwater has been degraded. In some coastal areas, intensive pumping of fresh groundwater has caused salt water to intrude into freshwater aquifers. Municipal and industrial wastes, chemical fertilizers, herbicides, and pesticides have entered the soil and infiltrated some aquifers.

Pesticides, nitrates, gasoline, mining residues, bacteria, and organic chemicals have all been discovered in well water. Some major pollution sources include

- leaking underground petroleum pipes and storage tanks
- failing septic tanks and improperly treated municipal sewage
- solid, liquid, or hazardous wastes improperly disposed of in landfills and other areas
- chemicals leaching into the soil from poorly-designed landfill sites
- over-application of fertilizers and pesticides; and improperly managed animal wastes.

Once contaminated, groundwater is difficult, expensive, and sometimes impossible to clean up. Groundwater contaminants move slowly and do not spread or mix quickly. Instead, they move in featherlike masses that can be present for years before being detected in drinking water. Cooperation between industry, agriculture, state and local governments, and individual citizens is essential if we are to protect our groundwater resources.

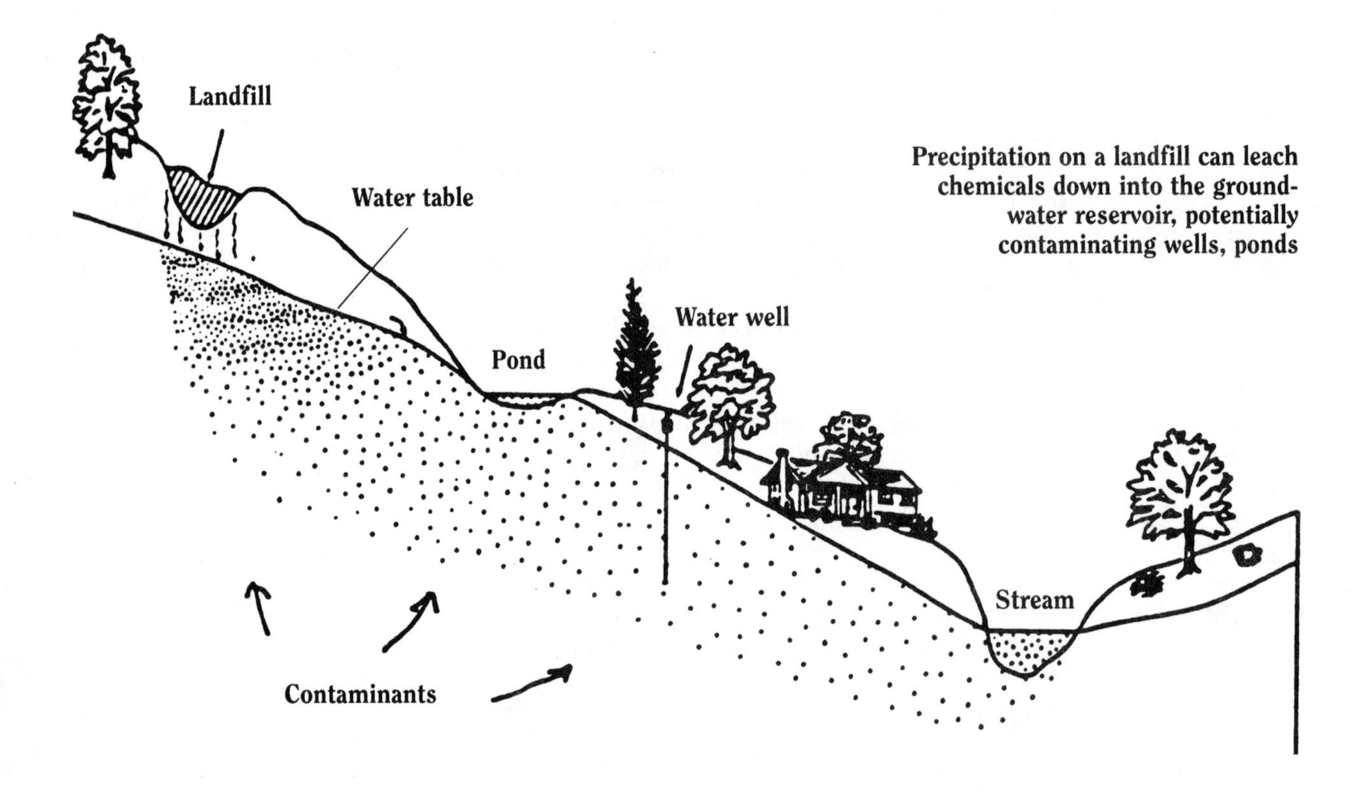

Precipitation on a landfill can leach chemicals down into the groundwater reservoir, potentially contaminating wells, ponds

Adapted from the U.S. Department of the Interior/Geological Survey booklet "Ground Water" (U.S. Government Printing Office: 1986-491-402/04), and "Groundwater Protection Begins at the Source," a booklet prepared by the Tennessee Valley Authority (TVA-ONRED 87/18).

READING 4

Managing Hands-on Activities in the Classroom

Students love laboratory activities! They enjoy using unfamiliar, exotic equipment, solving real problems, making decisions, and exchanging ideas with other students. Students in the middle grades view laboratory activities as an independent, grown-up way of learning. There is no better way of turning your students on to science.

Using laboratory activities to teach scientific concepts requires careful planning and organization if your students are to make optimal use of laboratory time. Planning is especially important if your students have had little hands-on experience previously. The following suggestions may help students succeed in hands-on activities, and make the teacher's task of classroom management easier.

◆Setting laboratory policies

Decide on your laboratory policies and expectations before your class performs its first laboratory activity. You have already been successful teaching students in a variety of learning environments; introduce hands-on activities with confidence!

Whatever policies you decide on, be sure that your students understand the consequences if they do not meet your expectations. Spending a small amount of time explaining your laboratory rules will improve your students' performance. Many teachers find it useful to post a formal list of laboratory rules in their classrooms. Others give the students a copy of the rules to read, discuss with their parents, and keep for future reference.

The following discussion of laboratory management issues may help you refine your own set of rules:

- **Laboratory safety:** Performing any experiment involves some risk of personal injury. Such risks can never be eliminated completely. However, if students understand and follow established safety guidelines, the risk of serious injuries can be greatly reduced.
- **Lab station assignments:** You may decide to assign workstations and lab partners before beginning hands-on activities. If so, select the members of each lab group using your knowledge of the academic ability and social dynamics of your students. Avoid groups larger than four students if you have enough equipment; groups of two or three students are often preferable.
- **Talking:** Students should be under control during laboratory sessions, but overzealous attempts to restrict talking may be counterproductive. Students can learn a great deal by discussing their observations and inferences with their lab partners while doing an experiment.
- **Equipment:** Maintaining science equipment in good operating condition is essential; damaged equipment may be dangerous or difficult to use. In some schools, students are responsible for the repair or replacement of materials damaged by careless or improper use. Remind your students to ask for help if they do not understand how to use a particular piece of equipment.
- **Class dismissal:** Budgeting about five minutes for students to clean and put away laboratory materials in an orderly fashion encourages responsible student behavior. A hurried cleanup will leave your room in a mess, make some students late for their next class, and leave you feeling harried and unprepared for your next group of students.

No single set of laboratory rules can meet the needs of all teachers for all science classes. Each teacher must decide on laboratory rules that are compatible with his or her educational philosophy and teaching style, and that optimize students' learning opportunities.

If you prepare well and start hands-on activities with confidence, your actions will say, "This laboratory activity is both important and fun, and I expect significant learning to take place." Discipline problems should be minimal. However, we recommend that you plan how you will deal with uncooperative students in advance of each laboratory activity. The specific disciplinary actions you take depend on your personal teaching style.

◆Preparing for hands-on activities

Choose brief, uncomplicated experiments with clear instructions for your students' first hands-on activities. As they gain experience, you will be able to increase the length of the activities.

A day or two before the activity, check and prepare the materials and equipment that you will need. You may find that you need to make repairs or find replacement materials.

Before distributing equipment to students, review the procedures for the activity with them. Since you will have tried out most of these activities in these modules during the teacher workshops, you will be able to alert students to any steps that may cause technical difficulties.

Take time to clarify your expectations for behavior during the activity, and state the consequences of uncooperative behavior. Encourage students to stay in or near their seats while performing the activity. When students demonstrate their ability to work independently, allow them more freedom *and* more responsibility during hands-on activities.

◆Performing hands-on activities

Depending on the arrangement of your classroom, you may decide to place laboratory equipment at individual lab stations or distribute it from a central location. In either case, be sure students understand that they are responsible for using the equipment properly, cleaning it, and returning it to its storage space as directed.

You can circulate, monitor, assist (but not too much), and ask questions while students are performing an activity. Enjoy watching the methods of science lead to "aha!" discoveries for your students. If a behavior problem arises, act quickly in accordance with your lab rules. Students need to witness that you will enforce your policy to protect the learning environment for all.

◆Following up hands-on activities

Discuss the activity with your class. Relate it to what they have learned in the past and what they will learn in the future. Encourage students to discuss their observations with other members of their group, their friends, and their families. Such discussions often produce interesting facts, suggestions, and questions to share with their classmates.

Writing about their laboratory experiences helps students understand the scientific concepts used in each activity. There are several ways to encourage students to write about activities:

- Each student can answer the activity questions while they are performing the experiment.
- Students can prepare a group lab report.
- Each student can write a paragraph explaining an activity in his or her own words. Writing assignments often help students discover gaps in their knowledge and may lead to some exciting class discussions!

READING 5

Flowing Water Reshapes the Earth

The greatest agent for shaping the surface of the Earth is moving water. Few landscapes escape the changes caused by water flowing from higher to lower elevations. When water follows the same set of connecting channels over a period of time, a river system is born.

Pictures taken by satellite show that Earth's river systems have tree-like patterns. When using this "tree analogy" to describe the Mississippi River system, the base of the "tree" is the river mouth, the place where the Mississippi enters the Gulf of Mexico. The main channel of the Mississippi River is the trunk; its main tributaries such as the Missouri and Ohio Rivers are large branches; the Platt and Monongahela Rivers are

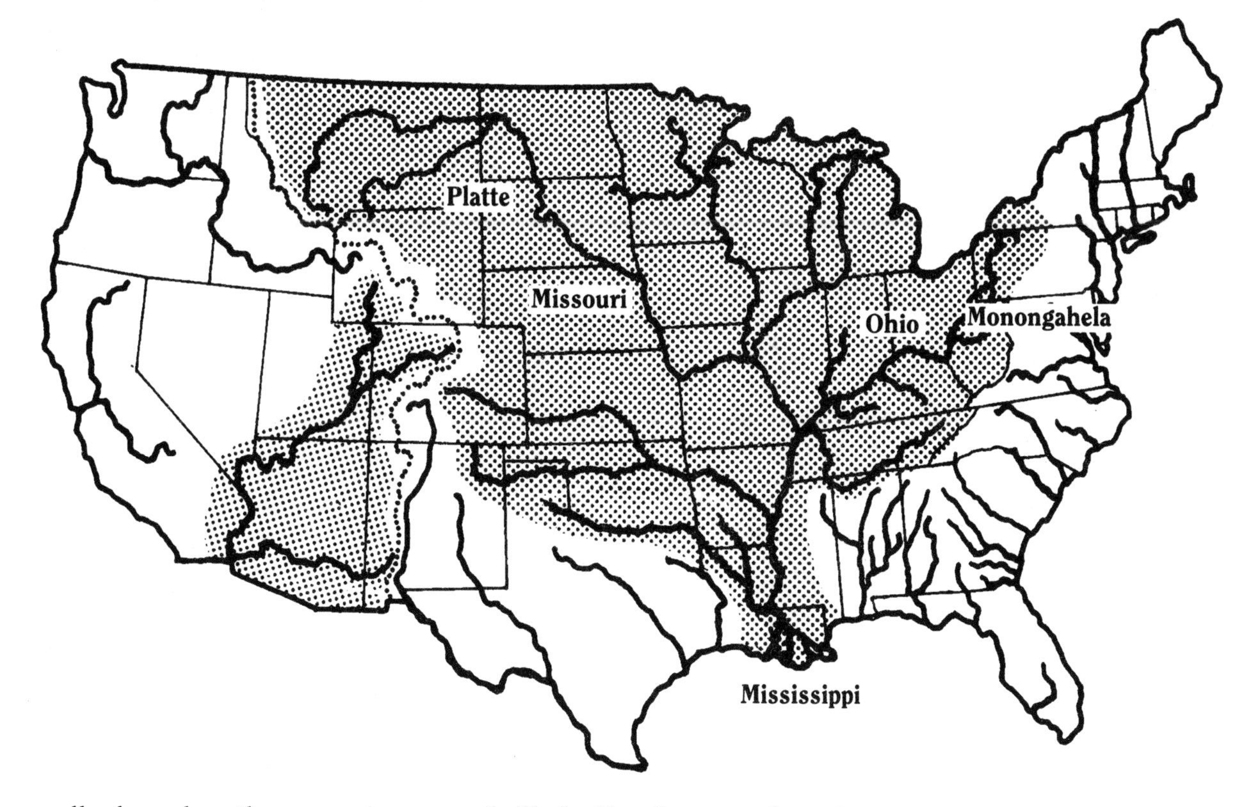

smaller branches; the many streams and rills feeding them are the twigs.

All of the area drained by a river system is called a **drainage basin**. One drainage basin is separated from another by a ridge or high area of land called a **divide**. The Continental Divide of the Rocky Mountains separates the Mississippi River and Colorado River drainage basins.

River systems become larger as erosion makes their numerous channels deeper, wider, and longer. The ability of a river to erode its channel is determined largely by

- the volume of water flowing in its channel, and
- the velocity of the water. Large water volumes flowing down steep slopes produce high velocities, and result in rapid erosion.

The size and type of sediments being carried by a river also affect the amount of erosion it causes. Velocity determines what size sediments a river can transport. Slow-moving streams can transport only small particles, but the sediment load of large, fast rivers may even include boulders and cobbles. These quickly erode the river's channel. Large sediments carried by fast-moving streams are not essential for reshaping a streambed, however; given a long period of time, small but highly abrasive sand and silt particles can cause a great deal of erosion.

Rivers erode their channels in different ways. Water tends to follow the shortest, most direct course downward, but when a stream encounters resistant, hard-to-erode rock, the water may cut a path around the obstacle, changing the course of the channel. On the other hand, if a river cuts slowly into resistant rock, a deep, narrow "box canyon" will be formed. If soil forming the river channel is relatively soft, the material along the sides of the channel will be undercut by the water and "slump." Soil falling into the channel will be carried away by the river.

A river's channel is constantly being deepened by erosion, but it is simultaneously being filled in as the river water deposits some of its sediment load in the channel. When a river's rate of sediment deposition is equal to its rate of erosion, the river reaches a state of equilibrium. This state will remain stable as long as (1) the elevation of the drainage basin remains the same, (2) the climate does not change, and (3) sea level does not increase or decrease. The state of equilibrium would also change if one drainage basin were "captured" by another through the erosion of a divide separating the basins.

Although rates are highly variable, under ideal conditions running water carries away about one centimeter of soil in 300 years. Most of this sediment does not reach the ocean directly but is temporarily stored within the drainage basin. As the water volume and velocity of a river fluctuate, sediment may be deposited in the channel, on either side of the channel, or at its mouth. The fine sediments spread out in the drainage basin on either side of the channel form a **floodplain**. Where seas are relatively calm, sediments deposited at the mouth of a river may form a fan or **delta**. The deltas of large rivers such as the Mississippi extend many kilometers into the open sea.

Rivers can be places of great beauty and serenity one moment and the scene of violence and destruction the next. The fate of a river can determine the fate of a culture. Ancient civilizations blossomed where flooding rivers left great gifts of fertile sediment, yet floods kill thousands of people and destroy millions of dollars worth of property each year.

READING 6

Rivers of Ice: Glaciers

During the Great Ice Age (the Pleistocene Epoch of the geologic time scale), mountain glaciers formed on all continents. The icecaps of Antarctica and Greenland were more extensive and thicker than today. Vast glaciers, in places 1500 m (5000 ft) thick, spread across northern North America and Eurasia. Almost a third of the present land surface of the Earth was intermittently covered by ice. So much water was locked in the ice of glaciers that sea level dropped 120 m (400 ft) lower than it is at present. Even today, remnants of these great glaciers cover almost a tenth of the land on our planet.

Although the Great Ice Age began a million or more years ago, the last major ice sheet to spread across the northern United States reached its maximum extent about 20,000 years ago. It lingered in Canada until about 6000 years ago, when most melted. Mountain glaciers and ice fields are the only remnants of the great glaciers on the mainland of North America.

We also know that glaciations occurred before the Great Ice Age. Scientists theorize that Earth's climate has fluctuated many times during the long course of geologic history, and that during the periods when the climate was cooler, large glaciers formed.

There is ample geological evidence indicating that large parts of the north-central, northwestern and northeastern United States were once overridden by thick ice sheets. In these areas, telltale traces of glacial erosion and deposition are widespread. Rock outcrops are smooth and polished, and/or they are scratched and striated; hills are rounded and covered by glacial debris; and valleys are choked by sand and gravel deposited by glacial melt waters.

All evidence indicates that slowly advancing ice sheets plowed up the soil and loose rock, plucked and gouged boulders from outcrops, and carried this material forward, often for great distances. Huge boulders called **erratics** (Latin for "wanderer") remain on glaciated landscapes, reminders of the earthmoving power of glaciers. Using the material in transit as an abrasive, the glaciers polished and scraped the rock outcrops, producing a smoother landscape. The great glaciers commonly softened the contours of the land by wearing down the tops of hills and filling the valleys. As the glaciers melted, unsorted mixtures of clay, sand, gravel, and boulders were deposited on the countryside.

States having glaciers	Approximate number
Alaska	thousands
Washington	950
California	290
Montana	200
Wyoming	100
Oregon	60
Colorado	25
Idaho	20
Nevada	5
Utah	1

Today we find glaciers in areas where, over a number of years, more snow falls than melts. As this snow accumulates and becomes thicker, it is compressed and changed into dense, solid ice. The ice in a glacier flows from the area of surplus snow accumulation to the area where yearly melting exceeds accumulation. The snow and ice of glaciers tends to flow due to its own weight—downhill if it is on a slope or out in all directions from the center if it is on a flat area.

On land, glaciers come to an end at places where the rate of icemelt equals the rate of iceflow. The situation is different for coastal glaciers that extend out into the ocean. Large chunks of ice break off of these glaciers and float away. We call these chunks of glacial ice **icebergs.**

Large areas of North America are still covered by glaciers. The largest ice field in North America is the Columbia Ice Field. Located on the border of Alberta and British Columbia, this field has several large glaciers issuing from it. It is estimated that 74,400 square km of Alaska (28,842 square miles, an area larger than the state of West Virginia) is covered by glaciers. Alaska's Bering Glacier, the largest glacier in the United States, is 203 km (126 miles) long and covers an area of more than 5000 square km (1900 square miles, an area almost the size of the state of Delaware).

In the lower 48 states, about 1650 glaciers (most of them very small) cover a total area of about 587 square km (227 square miles)—less than 1% of the total area of Alaskan glaciers.

Adapted from U.S. Department of the Interior/Geological Survey pamphlets "The Great Ice Age" by Louis L. Ray (U.S. Government Printing Office: 1983-381-618/16), and "Glaciers: A Water Resource" by Mark Meier and Austin Post (U.S. Government Printing Office: 1984-421-618/115).

READING 7

Danger! Rising Water!

Building lots with unobstructed views of rivers, lakes, or oceans are considered prime locations for construction of homes and commercial properties. Americans are willing to pay high prices for waterfront property. But many owners of these coveted properties with a view are in danger of losing the use of their land because of rising waters. Large areas of developed land are being affected by periodic flooding and/or severe erosion. Sea coast developments, riverfront properties, and even desert areas are being covered by water as lake and ocean water levels rise or as the land subsides.

In the late 1970s Great Salt Lake, a briny lake in the desert region of western Utah, began growing in size as a result of greater-than-average runoff from unusually heavy snowfall in nearby mountains. Between 1977 and 1987, the volume of water contained by the lake almost doubled. The lake's level (as of June 1987) is 12 feet higher than normal—the highest level recorded in the 140 years since the Mormons migrated to Utah.

By the mid 1980s the total economic damage to the region amounted to over $400 million. The expanding Great Salt Lake was engulfing ever-larger areas of farmland, periodically flooding highways leading into Salt Lake City, and endangering wildlife sanctuaries.

In an effort to prevent additional damage, the Utah legislature has spent over $60 million to build a huge evaporation pond in the desert to the west of the lake. Beginning in 1988, the excess lakewater is being pumped into the pond at a rate of 1.3 million gallons per minute. Theoretically, this will return the lake to normal levels, but there is no guarantee that this tremendous engineering gamble will work. If summers are cooler than average in the next few years (thus reducing the rate of evaporation), or if wet weather patterns intensify, the multimillion dollar pumping project will fail.

Eroding coastlines along the Atlantic and Pacific Oceans have made stories about houses falling into the ocean a regular feature of the evening news. Recently, even the landlocked Midwest has begun to experience the same type of problem: Lakefront homes and businesses in Chicago and other Great Lakes cities are in danger of being flooded. The levels of the Great Lakes are rising; their beaches are eroding. There is no obvious way to prevent increasing amounts of damage if current climatic patterns continue.

Potential changes in global climate threaten to dramatically change coastal geography the world over. The average temperature of the Earth is slowly increasing due to the greenhouse effect. The increasing concentration of carbon dioxide in the atmosphere caused by the burning of fossil fuels (coal, oil, and natural gas) increases the atmosphere's ability to hold heat. Heat is held in Earth's atmosphere in much the same way that a greenhouse traps heat and water inside.

If Earth's average temperature continues rising, polar ice masses will melt, raising the sea level. A rise of only 3 m in sea level would submerge much of the state of Florida. If one half of the water locked in the polar ice caps melted, sea level would rise by almost 30 m. Many of the world's great cities, including Washington, D.C. (presently 7.6 m above sea level), and Tokyo, Japan (elevation 9.1 m), would disappear under water. Berlin, Germany, presently at an elevation of 33.5 m above sea level, might become a seaport.

Sea level changes occur slowly. At present, scientists estimate that sea levels are increasing at a rate of 5 to 30 cm per century. At that rate, inundating Florida with 3 m of water would require at least 10 centuries—1000 years. However, no one knows for sure whether or not these rates will change. The time required to raise sea levels by several meters could be much shorter or longer than our current estimates. Even at present-day levels, flooding is a serious problem in many coastal areas.

The following news article (published September 19, 1987, and reprinted with the permission of the Scripps Howard News Service) gives a summary of a study done by the National Academy of Sciences on the types of problems that rising water may cause.

Sea Threatens Coastal Building, Academy of Sciences Panel Says

Robert Engelman
Scripps Howard service

WASHINGTON—Rising sea levels threaten to wash away new coastal construction over the next century if developers don't plan for them, a National Academy of Sciences panel said Friday.

"An inappropriate choice [in locating coastal buildings] could be very expensive," said a committee of coastal engineers and geologists studying the impact of expected sea level changes on coastal construction.

Builders of any structure expected to last 50 years or more should be particularly wary of assumptions that shorelines will stay put, the panel said.

But the committee also concluded that "there does not now seem to be reason for emergency action" and suggested that small structures like beachfront cottages are not much endangered by long-term sea rise.

"Sea level change during a structure's design service life should be considered along with other factors," the report concluded.

Sea levels have been rising slowly since the Ice Age ended about 18,000 years ago, according to the report, and they moved up between 4 and 12 inches on average over the last century. Each inch of horizontal elevation can drive the ocean eight feet inland on fairly level sandy ocean beaches.

The change in sea level is uneven, however, since coastal land often moves up or down irrespective of sea level.

In Louisiana, for example, recent extraction of groundwater and minerals along with wetlands destruction combine to push the land around the Mississippi River delta down. In parts of the state, described by the NAS panel as the U.S. coastal land most vulnerable to sea level rise, the ocean is moving 10 feet or more inland each year.

Alaska and northern Canada, by contrast, are still rebounding from the lifting of the load of mile-deep glaciers at the end of the Ice Age. Relative sea levels have actually dropped in those areas.

The pace of the sea level rise overall is expected to accelerate over the next century, however, as the Earth warms because of human activities. This "global warming" is predicted to result from increases in various gases—emitted by industrial processes, automobiles, and agriculture—that tend to lock the sun's heat in the atmosphere.

A warmer Earth would raise sea levels because melting polar ice would release water into the oceans and because warm water expands.

The likelihood of rapidly rising oceans means that coastal developers face three options, the NAS panel said: fortify shores with seawalls, jetties, and levees; manage them by replacing depleted sand; or "let nature take its course" by building or moving structures well back from the shoreline.

Jetties and seawalls only shift destructive erosion to nearby beaches, while sand management is an expensive stopgap measure, the panel concluded. The scientists called instead for retreat from shorelines "in a planned and orderly fashion," noting that some state policies already restrict new construction on seashores.

Whatever approach developers use, permanent installations are likely to feel the impact of rising seas—from flooded airport runways, to bridges too close to the water for boats to pass underneath, to the contamination of wells and underground water supplies by salt water.

READING 8

Scientific Literacy for All

We live in an age of scientific and technological innovation; every phase of our lives is touched by the products and processes of these two enterprises. While pre-college preparation of future scientists is vitally important, training future scientists cannot be the sole function of a K-12 science program. Our technological society demands that *all* future citizens be scientifically literate.

As our increasingly sophisticated technology provides us with new industrial capabilities and more personal leisure time, it also confronts us with new sets of problems. Hazardous waste, acid rain, mineral exploration, water rights—these topics appear in the news daily, and are campaign issues for many politicians. Important decisions about the allocation of Earth's resources are being made all the time. Citizens must have a working knowledge of science and mathematics in order to participate intelligently and responsibly in the decision-making process.

Several national groups and numerous individuals have attempted to define scientific literacy. These definitions are useful because they give direction to those who teach and develop instructional materials. Several of these definitions have analyzed scientific literacy in terms of the behaviors desired of a scientifically literate person. For example, according to one definition,* the scientifically literate person

- has knowledge of the major concepts, principles, laws, and theories of science and applies these in appropriate ways
- uses the processes of science in solving problems, making decisions, and other suitable ways
- understands the nature of science and the scientific enterprise
- understands the partnership of science and technology and its interaction with society
- has developed science-related skills that enable him or her to function effectively in careers, leisure activities, and other roles
- possesses attitudes and values that are in harmony with those of science and a free society
- has developed interests that will lead to a richer and more satisfying life, and a life that will include science and lifelong learning

There are many ways to develop each of the above components of scientific literacy while teaching about water in the Earth's systems. A given activity, reading passage, or film often contributes to several components. It is not necessary to have students do one activity to teach them certain principles of science, another activity to teach processes of science, and another to show how science affects our society.

The activities in these modules use the processes of science to teach problem solving and to encourage the development of critical-thinking skills. For example, students performing the activity "How Can Farmers Reduce Erosion Caused by Rain?" use a scientific model to simulate a plowed hilly field, manipulate the variables of soil moisture and plowing technique, make observations of the model, and use the data collected to recommend ways of dealing with an important economic problem: preserving the fertility of our agricultural land.

As you use these activities with your students, you can also provide

---

*Simpson, Ronald D., and Norman D. Anderson, *Science, Students, and Schools: A Guide for the Middle and Secondary School Teacher.* New York: John Wiley and Sons, 1981.

information about the concepts being studied through written materials, films, lectures, and discussions. Together these reveal the nature of science: a dynamic partnership between knowledge and process.

◆Equity in education

Equity—equal access to educational, employment, and housing opportunities—is a familiar topic to most Americans, largely because of the dominance of civil rights issues in the past two decades. In the past, there were very few opportunities for women and minorities to have careers in science and science-related fields such as engineering and medicine. This is no longer true. However, these groups still encounter many barriers to full participation in science, often because of sex and race role stereotyping.

Researchers have found that when boys and girls are working together in groups, the boys will set up the apparatus and do the experiment while the girls take notes and record the results. In addition, teachers often ask boys harder questions than they ask girls, give boys more time to think about their answers, and in other ways communicate that they have higher expectations for boys than for girls. Instructional materials frequently ignore the contributions of women and minorities in science, or portray these groups in stereotyped ways, thus depriving female and minority students of appropriate role models.

Equity in science education means encouraging female and minority students, as well as white males, to participate actively in science lessons. Teaching styles that are supportive of an "equal opportunity" classroom become especially important in light of the changing demographics of the American populace. The composition of U.S. citizenry is shifting. The birth rate of the American white population is declining, whereas that of the nonwhite population (particularly Hispanic Americans) is increasing rapidly. "Minorities" are rapidly becoming the majority of the student population in many places. It is, therefore, even more important that all American youth receive the very best education available.

Educational research has begun to identify some of the components of effective and equitable teaching. For example, during the past decade researchers have learned that there are many beneficial outcomes when students work in cooperative groups on a task with each person contributing to the learning of the others. Most importantly, non-competitive methods seem to make learning more accessible to a wide range of students. Advanced students develop their leadership skills while they reinforce their subject knowledge. Students who learn at a less rapid rate have time to discover answers via trial and error and to observe the logical processes demonstrated by their peers. In general, cooperative learning activities encourage students to develop learning skills by providing a supportive climate where students can learn with less risk of embarassment or failure.

Knowing how to determine the pH of an acidic solution being used in an experiment is an important skill for future scientists, but understanding the scientific and political implications of acid rain, conserving water supplies, and using scientific knowledge while participating in the democratic decision-making process are important tasks for *all* citizens. Group activities allow students of widely varying backgrounds and abilities to address such issues in their science classes.

No one advocates that all lessons be conducted in this way; there is a place for individual work, for drill and practice, for whole-group discussions, and for other teaching and learning methods. However, cooperative group learning is particularly well-suited for use in laboratory settings and for carrying out some of the other activities described in these materials.

Students who have spent their school years in classrooms where cooperation was not encouraged or even allowed will not be able to work cooperatively without training in this kind of group work. It takes time and practice to learn to work interdependently. If you are interested in learning more about the positive outcomes produced by this kind of teaching, you may wish to read *Circles of Learning: Cooperation in the Classroom* by D. W. Johnson and R. T. Johnson (Alexandria, Va.: Association for Supervision and Curriculum Development, 1984).

◆Water and careers

For some individuals, the topic of water resources is more than a voting issue; it presents an employment opportunity. Jobs in surface and groundwater management are projected as one of the fastest-growing employment areas (Department of Labor Statistics, 1986). In private industry, firms that address hazardous waste and water-quality problems usually have several hydrologists on their staffs. Government agencies such as the Environmental Protection Agency and the Departments of the Interior and Agriculture are some of the primary employers of water-resource personnel.

The following reading, "Water and the Jobs of Your Lifetime," is intended for student use. It briefly describes how our industrial society uses its water supplies and also points out some of the relationships between water resources and individual employment opportunities. You might wish to have students clip news and magazine articles on water issues: for example, quality, flooding, scarcity, treatment, and erosion, to heighten their awareness of local, state, national, and international industries and the related jobs.

READING 9

Water and the Jobs of Your Lifetime

◆Water and life

Planet Earth is unique in our solar system because of its abundant, available water. Life developed on this planet within the protective environment water offers, and water is essential to the functioning of living things. Your body is over 70% water. Your existence and the quality of your life are totally dependent on water.

"Water-intensive" could be substituted for the phrase "energy-intensive" often used to describe our society. Human beings require less than 2 L (about 1/2 gallon) of water per day to survive, but in the United States, the average person uses 340 L (about 90 gallons) each day at home. U.S. industries use over 1270 billion L of water a day—a volume greater than half the average daily flow of the Mississippi River at Vicksburg, Mississippi. The combined industrial, agricultural, and personal use of water in the United States amounts to approximately 6000 L (1600 gallons) per person per day. Let's examine the relationships between some occupations and the production and use of this vast amount of water.

◆Providing clean water

Getting water from the ground or the surface to locations where it can be used for drinking involves a tremendous variety of jobs. *Water treatment plant operators* control and maintain equipment, regulate flow, collect samples, perform laboratory tests, keep records, and make repairs to equipment. *Water meter readers* record water usage, inspect meters, and may install or repair them. *Well drillers* tap groundwater supplies where adequate surface water is not available. The sale, installation, and maintenance of water-system hardware—pipes, pumps, filtering systems, monitoring systems, etc.—require workers who can solve practical problems, communicate with others, and work with their hands.

Geologists, scientists who study the Earth, are often called upon to provide information and advice about obtaining and protecting water supplies. Depending on their training and interests, geologists may work outdoors, in the laboratory, or in an office. *Hydrogeologists* specialize in studying the occurrence, movement, quantity, and quality of water under the ground and on the Earth's surface (including snow and ice). They help develop, control, and protect water supplies, and assist farmers with irrigation needs. They may work with *environmental geologists* to solve problems of water pollution, chemical or radioactive waste disposal, flooding, or erosion.

◆Growing your food

Agriculture is a big business in the United States. Growing crops requires water, and lots of it, but farmers are not the only people involved in food production whose jobs depend on adequate water supplies. Water is essential in *all* the steps of getting food to your table. Even before seeds are sold to the farmer, *agricultural researchers* need water for their work of developing healthy productive plants and determining the best soil and water conditions for growing them. *Agricultural engineers* design irrigation systems, flood control projects and farm machinery for varying water conditions. After the crops are harvested, *food processors* use large amounts of water to prepare the food for sale to the consumer.

◆Building your shelter

Water is essential for producing the materials used to build our homes. Trees, like any other crop, require water to grow. The manufacturing processes for converting logs into finished lumber and other wood products consume a great deal of water. Since our forest resources are being rapidly depleted, *research foresters* are working to increase forest productivity by studying trees' water and nutrient intake, and by using genetic techniques to develop faster-growing trees.

In addition to lumber, your home contains materials made from natural resources that come from the Earth. Its concrete foundation, copper wires, iron pipes, and chrome faucets all come from mineral resources located by geologists. Once mineral resources are located, an *economic geologist* evaluates their profitability and assists in their development. Processing these materials consumes enormous amounts of water.

A *mining geologist* will be in charge of the "down-to-earth" work of geology, removing the resources from the Earth. The knowledge and skills of an *environmental geologist* will be necessary if the mining and the processing produce waste by-products capable of polluting nearby streams and aquifers (layers of rock or sediment saturated with water).

Engineering geologists work on large construction projects to prevent possible damage from floods, earthquakes, landslides, coastal erosion, or other natural hazards. Working along with and supporting all of these efforts to create your home are large numbers of specialty workers such as *miners, carpenters, plumbers,* and *electricians*. All of the products, processes, and workers employed in construction require lots of readily available water.

◆Keeping you warm

In most areas, your house alone will not keep you warm in the winter. Obtaining the fuel that does this job usually involves water somewhere along its path of development and transportation. The easy-to-reach oil supplies on land are dwindling, and *marine geologists* are searching for oil and natural gas under the oceans. They are also investigating mineral resources of the continental shelves. *Geophysicists* are looking for geothermal (hot water) energy close enough to the surface to be used as heat directly, or to generate electricity.

In some homes, electricity is used for heat as well as for light. Electricity can be produced by water that falls on and moves a turbine, or by water that is converted to steam to drive the turbine. Both nuclear-powered and conventional (coal or oil-fired) power plants use large volumes of water for cooling. Understanding the flow of water and its potential for producing energy is the job of the *hydraulic engineer*, who in turn depends on *mechanics, electricians, plumbers,* and other technicians to keep the complicated machinery humming.

Heating your home keeps you warm when you are inside. But what protects you from the cold when you are outdoors? Clothing, of course, and sometimes several layers of it. Whether we use cotton (plant product), wool (an animal product), nylon (a synthetic material), or something else, lots of water is used from the beginning to the end product. Growing, spinning, weaving, dying, and sewing are all water-intensive. Just think for a moment of all the jobs—and all of the water—necessary to produce that warm jacket of yours!

◆Recreation

Food and shelter are not enough to keep people happy. We are unique in our ability to think, remember, plan, dream, and have fun! And, believe it or not, even these activities require water. Producing all of those skis, skateboards, tennis racquets, and windsurfers, as well as the specialized clothing and shoes you need, requires water. When you plan an outdoor activity, you rely on the *meteorologist* to keep you informed as to what kind of weather to expect. We keep returning to water, this time as rain, snow, sleet, or hail!

One of the best ways to keep all of the muscles of your body fit is to exercise them in water. For such activities you will need the special instruction offered by a swimming or *water aerobics instructor*. Water is also used by a *physical therapist* to help exercise and strengthen muscles that have been injured.

◆NOW is the time to plan

As you can see, all aspects of work in the United States—from mining and manufacturing to cattle raising and farming, operating chemical companies, developing real estate, drilling for petroleum, producing electricity, running small businesses, and providing recreational and educational activities—require getting and maintaining an adequate water supply. Many large corporations have a full-time water expert on their staff, and most others hire water consultants from time to time.

Science and technology influence most aspects of our complex industrial society. You can keep more career options open by studying as much science and mathematics as possible. Graduating from high school is a minimum requirement for almost any satisfying job. Some water-related careers require a college education with an emphasis in the Earth sciences. Most others require a high school diploma followed by specialized vocational education and/or on-the-job training. Regardless of what career you choose, water supplies will play an important role.

READING 10

The Invisible Gas Becomes Visible: Clouds

Atmospheric air contains water vapor—water in its gaseous state. Most of the water vapor gets into the air across the interface between ocean and air. Nevertheless, a great deal of water is evaporated over land as well; transpiration from plants is a major source of moisture in the atmosphere.

We commonly think of the amount of moisture in the air in terms of relative humidity. **Relative humidity** is the ratio of the amount of moisture in the air to the maximum amount of moisture the air could hold under the same conditions; it is usually expressed as a percent. For example, when the air contains half as much water as it is capable of holding, we say the relative humidity is 50%. When the air contains all the water it can possibly absorb, the relative humidity is 100%.

Relative humidity is a convenient measure of how close the air is to saturation with water vapor. It is also a simple indicator of human comfort. On warm days when the relative humidity is high, evaporation of perspiration from your skin takes place slowly. We might refer to these as sticky days.

Water vapor changes to liquid water in the air by condensation, the change of state that is just the opposite of evaporation. **Condensation** occurs when the relative humidity is near 100%, and the vapor has a solid surface on which to collect.

Small solid particles are always present in unfiltered air. When water vapor condenses on them, we call these particles **condensation surfaces.** A salt crystal left drifting in the air from ocean spray that has evaporated, a particle of ash or oil from the smoke of a fire, or dust particles may serve as condensation nuclei. Sea salt and smoke particles from burning are especially suited for condensation nuclei, since water vapor condenses on these particles at relative humidities less than 100%.

The temperature at which the air becomes saturated with water vapor is called the **dew point**. This is the temperature at which moisture begins to condense on solid surfaces, forming **dew**. When the temperature of a surface reaches freezing before saturation occurs, water vapor changes directly to ice and **frost** forms.

While a number of factors control evaporation of water into the atmosphere, the cause of condensation can be summed up in a single phrase: cooling of the air. Air masses are cooled as they rise and expand; at high elevations there is both less atmospheric pressure (allowing for expansion) and cooler temperatures. Therefore, rising air accounts for most of the condensation that occurs in the atmosphere.

This explains why clouds and rain occur most commonly on the windward side of mountain ranges. As moist air strikes the windward side of a mountain, it rises, expands, and cools. It then condenses on condensation nuclei, forming clouds.

Western Oregon and Washington State have very moist climates; the eastern areas of these states are very dry in comparison. Their climates are different because the clouds forming on the windward (ocean-facing) sides of the mountains in these states drop most of their moisture as rain near the Pacific coast. The conditions leading to cloud formation no longer exist as the moisture-poor air descends the leeward slope of the mountains, where it is warmed and compressed. The western sides of the mountains often receive a warm, dry, chinook wind rather than rain.

◆Why do clouds differ?

If the upward motion of a large air mass is slow and gradual, and condensation occurs throughout the air mass, a layer of cloud covering a wide area (clouds from horizon to horizon) will be produced. Under these conditions the sky is said to be overcast, and the clouds appear as sheets or layers covering a large portion of the sky. These are called *stratiform* or *stratus-type clouds*. They give the sky an overall gray appearance.

Sometimes clouds are well separated from each other and scattered over the sky with clear spaces between them. They may have a puffy appearance like a ball of cotton. These clouds are called *cumuliform* or *cumulus-type clouds*.

The sun does not heat the Earth's surface uniformly; heated "bubbles" of air form over warmer surfaces. Cumulus clouds form when bubbles of heated air rise through the surrounding air like helium-filled balloons. As they rise, the air expands and cools. If the air in the bubble cools to its dew point, condensation takes place; if it continues to rise, a cumulus cloud forms.

Clouds are composed of millions of water droplets or of ice crystals or of a combination of the two. It is sometimes difficult to determine the composition from the appearance of clouds. The thin wisps or faint sheets of clouds at high altitudes are usually made of ice crystals. If you have ever seen a halo around the sun or moon, it was probably caused by such a cloud. These ice crystal clouds are called *cirrostratus*. The white trails left across the sky by high-flying jet planes are true ice-crystal clouds. Ice-crystal clouds often have a fuzzy appearance at their edge in contrast to water-droplet clouds, which usually have sharp and clearly defined edges.

Clouds that form at the Earth's surface are called **fog**. Fog is usually caused by cooling of warm, moist air as it comes in contact with a cold land or water surface. The warm air loses heat to the cold ground or ocean surface below and cools to its dew point. Then the water vapor condenses to form fog.

◆Precipitation

The air's upward and downward motions, together with the supply of water, largely control condensation in the atmosphere. Cloud droplets are so small that the resistance of the air helps keep them from falling; if the water is to fall from the atmosphere, larger particles called raindrops or snowflakes must form. Raindrops are many times larger than cloud droplets and are pulled to Earth by gravity.

Snowflakes occur in beautiful hexagonal (six-sided) shapes. **Snowflakes** form when water vapor accumulates on ice crystals, going directly to the ice phase. **Sleet**, consisting of clear pellets of ice, is formed when raindrops fall through a layer of cold air and freeze.

Another type of frozen precipitation is **hail**, which usually falls from cumuliform clouds that have developed into cumulonimbus clouds or thunderheads. Hail consists of balls or irregular lumps of solid water. They are either transparent or composed of layers of ice and snow. The layers form as the hail is alternately lifted into a part of the cloud where snow and ice crystals become attached to it and then released into layers composed of water droplets. Hail the size of an orange is reported several times each year, usually in the midwestern states. Hailstones of this size and smaller ones, too, can do tremendous damage to crops, livestock, roofs, and automobiles and can seriously injure people.

This reading was adapted from pages 180–192 of *Investigating the Earth*—the Earth Science Curriculum Project (ESCP) of the American Geological Institute, ©1972, published by Houghton Mifflin Company, Boston, Massachusetts.

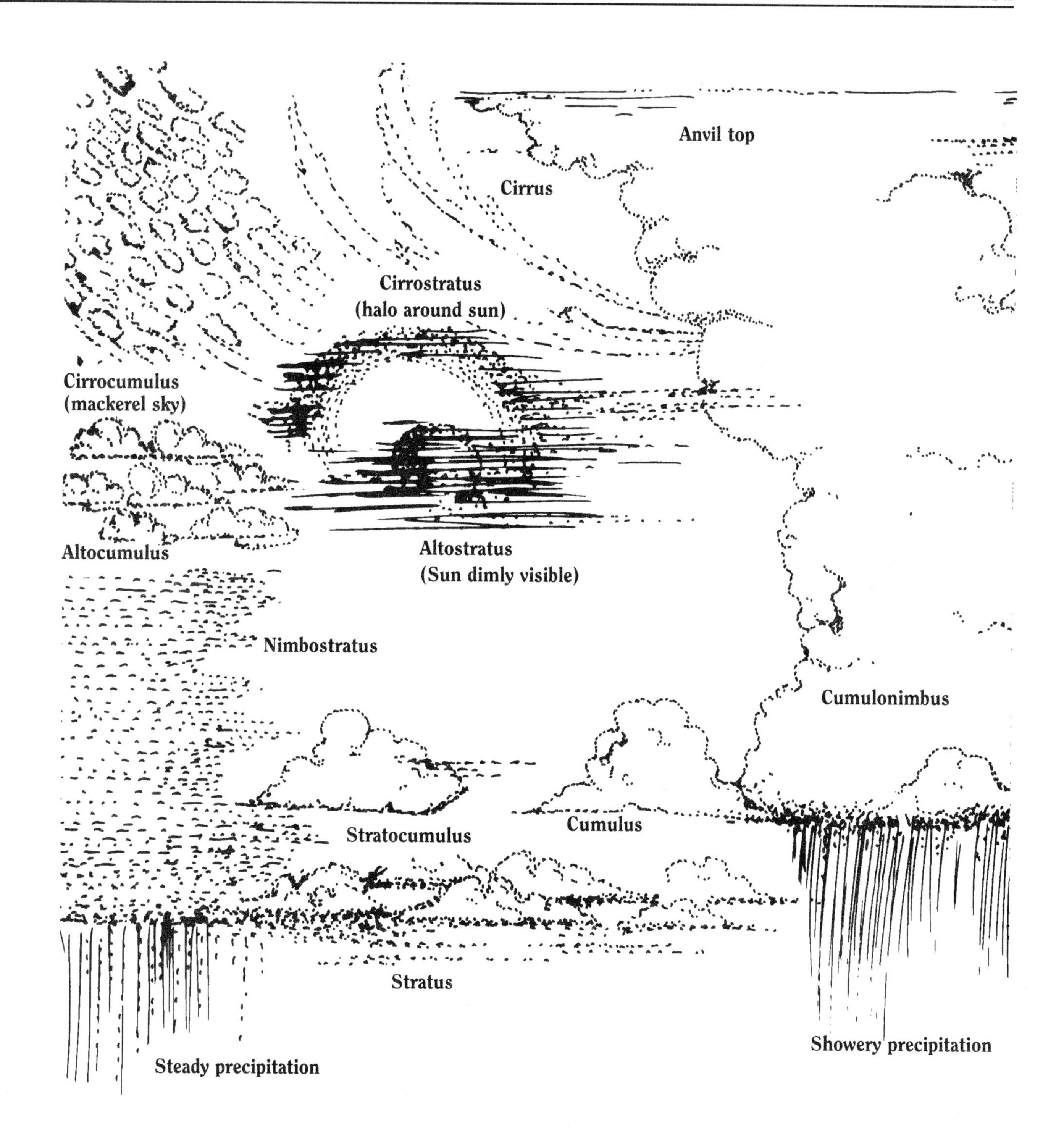
Anvil top
Cirrus
Cirrostratus
(halo around sun)
Cirrocumulus
(mackerel sky)
Altocumulus
Altostratus
(Sun dimly visible)
Nimbostratus
Cumulonimbus
Stratocumulus
Cumulus
Stratus
Showery precipitation
Steady precipitation

READING 11

How Acidic Is the Rain?

◆What is acid rain?

Acid rain, or acid precipitation, refers to any precipitation having a pH value less than that of normal rainwater. Carbon dioxide (CO_2) in the atmosphere makes normal rain slightly acidic. This is because carbon dioxide and water combine to form carbonic acid, commonly known as carbonated water. Its pH generally ranges from 5.0 to 5.6. Acid rain falling in the northeastern United States is often in the pH range of 4.0 to 4.6. Because "acid rain" includes snow, sleet, hail, dew, frost, and fog, it may also be referred to as acid precipitation or wet deposition. Some of the acids which fall as gases or solids are called dry deposition.

◆What is pH?

The pH scale, which ranges from 0 to 14, is used to measure whether a substance is acidic or basic. Pure water is neither acidic nor basic. It has a pH of 7, which is neutral. Values on the pH scale below 7 are acidic, and those above 7 are basic. Substances having a pH of 1 (battery acid, for example) are very acidic; substances having pH of 13 (such as lye) are very basic.

The pH scale is logarithmic, with a tenfold difference between each number. If the pH drops from 7 to 6, the acidity is 10 times greater; if it drops from 7 to 5, it is one hundred times greater; from 7 to 4, one thousand times greater and so on. For example, vinegar at pH 3 is 10,000 times more acidic than distilled (neutral) water.

◆What causes acid rain?

The burning of fossil fuels, particularly fuels such as coal or oil that contain sulfur, contributes to the acidity of rain. During burning, the sulfur in these fuels combines with oxygen from the atmosphere, forming sulfur dioxide (SO_2); burning also produces nitrogen oxides. The sulfur dioxide and nitrogen oxides are carried up the smokestack by the hot exhaust gases. They may remain mixed in the air for several days. The longer they remain in the air, the greater the likelihood that they will form solutions of sulfuric acid (H_2SO_4), and nitric acid (HNO_3). These acids may be dissolved in droplets of water and carried by winds for many miles.

The largest sources of sulfur dioxide and nitrogen oxides are coal-fired electric power generating stations. Industrial processes such as the smelting of sulfide minerals also contribute to the problem. Exhaust gases from cars and trucks also release large amounts of nitrogen oxides into the air, causing serious air pollution in high traffic areas.

◆Where is acid rain a threat to aquatic ecosystems?

In most regions of the United States, lakes and streams can tolerate some extra acidity with no adverse environmental consequences. This is true because the soil's alkalinity (typically from limestone naturally present in the soil) neutralizes the additional acidity.

However, some parts of the United States and eastern Canada are particularly sensitive to acidity. These regions have thin soils covering bedrock that is low in limestone content. Lakes and streams in these areas, lacking a naturally-occurring source of alkalinity, have little or no capacity to neutralize acid rain. Parts of the Rocky Mountains and the north-central, southeastern, and northwestern United States are among those susceptible to the effects of acid rain.

Pollutants that produce acid rain can be carried long distances by the wind. Industrial activity does not have to occur in an acid rain-sensitive area in order for that area to be affected. Watersheds in Canada are being damaged by acid rain produced by power plants in the United States. Acid rain is dissolving the limestone exteriors of buildings in areas far removed from major industrial activity.

◆How does acid rain affect water-dwelling animals?

Some aquatic organisms are more tolerant of acid conditions than others. Species living in naturally acid waters such as bogs are less likely to be sensitive to moderately low pH. However, many species of fish, such as rainbow trout, brown trout, smallmouth bass, and fathead minnows are unable to survive below pH 5.

Although adult forms of most aquatic species are generally more tolerant, the eggs and larvae are sensitive to low pH and are unable to survive. As the water becomes more acidic, fewer eggs hatch and fish may not grow to maturity.

Many species of amphibians (frogs, toads, and salamanders) breed in temporary pools that are formed by spring rains and melted snow. These pools may be very acidic. This acidity in these temporary pools may cause deformities and death in the eggs and developing embryos.

◆What can you do to help reduce acid rain?

We can all help by using fossil fuels more wisely. We can use carpools and mass transit; properly maintain our cars and trucks and their pollution-control devices; be more efficient in heating our homes; and conserve electricity by turning off lights and electrical appliances when they are not in use.

This reading contains excerpts from the poster "Acid Rain: The Effect on Aquatic Species" developed by the U.S. Department of the Interior, Fish and Wildlife Service, National Ecology Center—Leetown, Box 705, Kearneysville, WV 25430.

READING 12

Soil Erosion: The Work of Uncontrolled Water

When rain falls or water runs downhill on bare soil, it moves soil particles, organic matter, and soluble minerals. That is soil erosion.

The process of soil erosion by water consists of three principal steps: (1) loosening soil particles by the impact of raindrops or by the scouring action of runoff, (2) moving the detached particles by flowing water, and (3) depositing the particles at new locations. These steps occur in sequence from ridge to river in a **watershed**—a geographical area from which water drains to a single point.

Raindrops strike with enough force to tear clumps of unprotected soil apart and separate the tiny particles from each other. They splash the bits of soil about and gradually move them downhill.

Whenever rain falls faster than it can soak into the ground, a sheet of water collects on the surface and flows downhill. Falling raindrops continue to dislodge soil particles—keeping them suspended in moving sheets of water—or feed them into the little streams of water flowing off the field along crop rows or small gullies called **rills.**

Mineral nutrients and organic matter are churned into this **runoff** (precipitation not absorbed by the soil) and carried away, leaving the coarser, less fertile particles behind. The combined actions of beating rain and flowing water remove continuous layers of soil from fields. This is called **sheet erosion.**

If the water in these rills moves fast enough, it too dislodges soil particles that are carried along with those splashed up by the raindrops' impact. This scouring action carves out channels that join farther down the slopes. This is called **rill erosion.**

The rills carry more soil as they pick up speed or grow in size. The abrasive particles they carry help scour the sides and bottoms of the channels. This combination of sheet and rill erosion removes enormous amounts of soil from unprotected fields.

Soil erosion by water occurs anywhere there is enough rain to cause runoff, or where land is flooded by irrigation, snowmelt, or other causes.

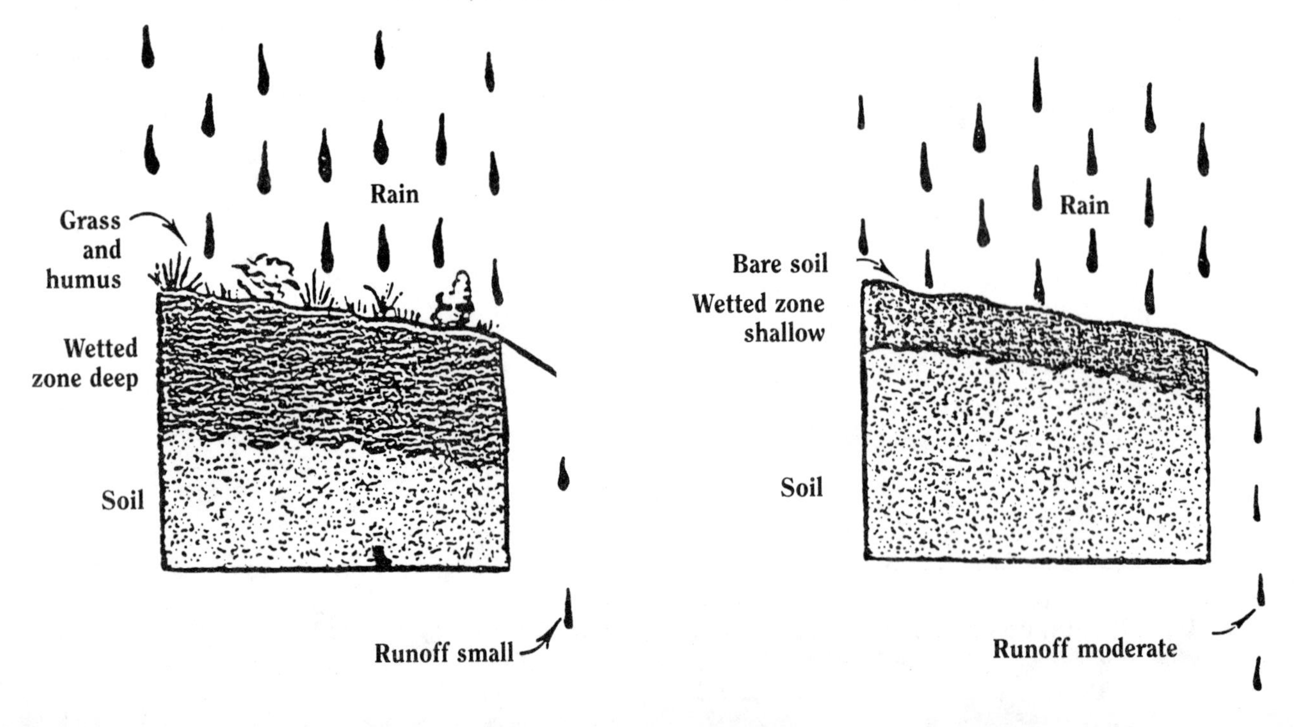

To avoid erosion, the soil must be protected from moving water. Dense vegetation—cover crops, mulches, grasses, or trees—will intercept rain and slow runoff. Barriers like terraces, sown strips of different crops, or plowing along the contours of hills can help control runoff where plowing leaves the soil exposed. The furrows produced by contour plowing act as long, curving dams that hold back runoff water. Otherwise, plowing has to be confined to nearly level soils where water moves slowly.

Land that is carelessly used for grazing or logging may also be damaged by erosion. For example, heavy grazing or improper cutting or burning practices can accelerate soil loss. However, with good management, soil cover can be maintained, and grasslands and woodlands can be protected from erosion.

Erosion by water has already damaged much of the farmland in the United States. Some soils, inherently unsuited for cultivation or so badly damaged they cannot be restored, need to be converted to other uses.

On cultivated soils, preventing excessive erosion is a major concern. Positive steps to improve the soil and use it efficiently are a part of modern soil conservation techniques. But without protection from erosion, soil improvement efforts fail. The seriousness of the problems caused by erosion depends largely on the kind of soil.

The small watershed is the natural land unit for controlling water and preventing erosion. What happens in each small watershed affects the land and people in the valley below.

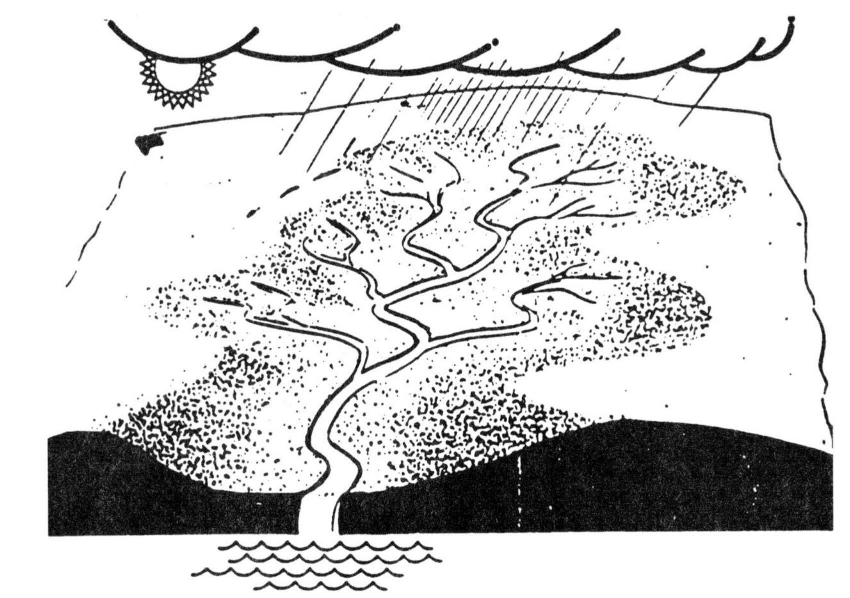

Erosion damage starts where the rain falls in the watershed. As runoff moves downhill, it grows in volume and in erosive force. Control is easiest if planned from ridge to river.

Flood-plain farmers are concerned with erosion on the uplands. Soil washed from the slopes may cover their crops. It may deposit infertile materials on good land and impair the drainage of their fields.

City dwellers, too, are affected by erosion. Silt chokes streams and fills reservoirs, it pollutes water supplies and depletes water storage, it destroys wildlife habitat and decreases recreational opportunities.

Unchecked runoff from big rains turns into floods that damage both farm and city property and endanger livestock and human lives.

Every soil is dynamic. It is constantly changing as water comes and goes and plants and animals live and die. Wind, water, ice, and gravity move soil particles around—sometimes rapidly, sometimes slowly. But even though a soil is never exactly the same for any measurable time, the layers in most soils stay much the same for many years unless they are changed by erosion. Soil erosion can remove, in a few years or even in a few hours, the surface layer of soil that was formed during hundreds or thousands of years.

The first European settlers in the United States were not faced with such problems—the land they arrived in was in equilibrium with the environment. But erosion speeded up as the settlers cleared away the vegetation to grow crops, cut timber, and let their livestock graze on the grasslands. It is the accelerated erosion caused by overworking the land that today's agriculturists are concerned with. Our goal is to keep the rate of erosion on cultivated land almost the same as the erosion that occurs on undisturbed areas of land.

Some soil conservation districts and municipalities sponsor erosion-control projects together. In this way, farm and city people can work together on watershed-protection and flood-prevention projects planned for entire watersheds.

Adapted from Agriculture Information Bulletin 260, "Soil Erosion—The Work of Uncontrolled Water," revised 1981. This booklet was produced by the Soil Conservation Service of the U.S. Department of Agriculture and is available through the U.S. Government Printing Office as publication 1981: 0-351-660.

READING 13

Conservation and the Water Cycle

◆How hydrologic processes affect the Earth and its inhabitants

The water cycle is an endless process of water circulation going on throughout the world.

To trace the movement of water through the cycle, begin at the far right of the diagram. There the sun's energy is transferring water from the sea and Earth to the atmosphere in the form of water vapor. The soil and inland water bodies through **evaporation** and plants through **transpiration** add large amounts of water vapor to the atmosphere, but most of it comes from the oceans. Man, animals, and machines add small amounts by means of **respiration** and **combustion.**

Air masses (top of diagram) carry the water vapor across the Earth, and the water vapor condenses into **precipitation.**

At the left, precipitation falls as rain, snow, sleet, and hail. Some evaporates while falling and returns to the atmosphere. A small amount is intercepted and held by plants or by buildings, automobiles, and other structures and machines until it evaporates back into the atmosphere.

Most of the precipitation soaks into the soil; the part that doesn't runs to the sea by way of streams and rivers. Groundwater gets there more slowly.

Misuse and poor management of the soil will decrease the amount of water that soaks into the soil and increase the amount that runs off over the surface. Runoff on bare land leads to erosion. Grass, trees, and other

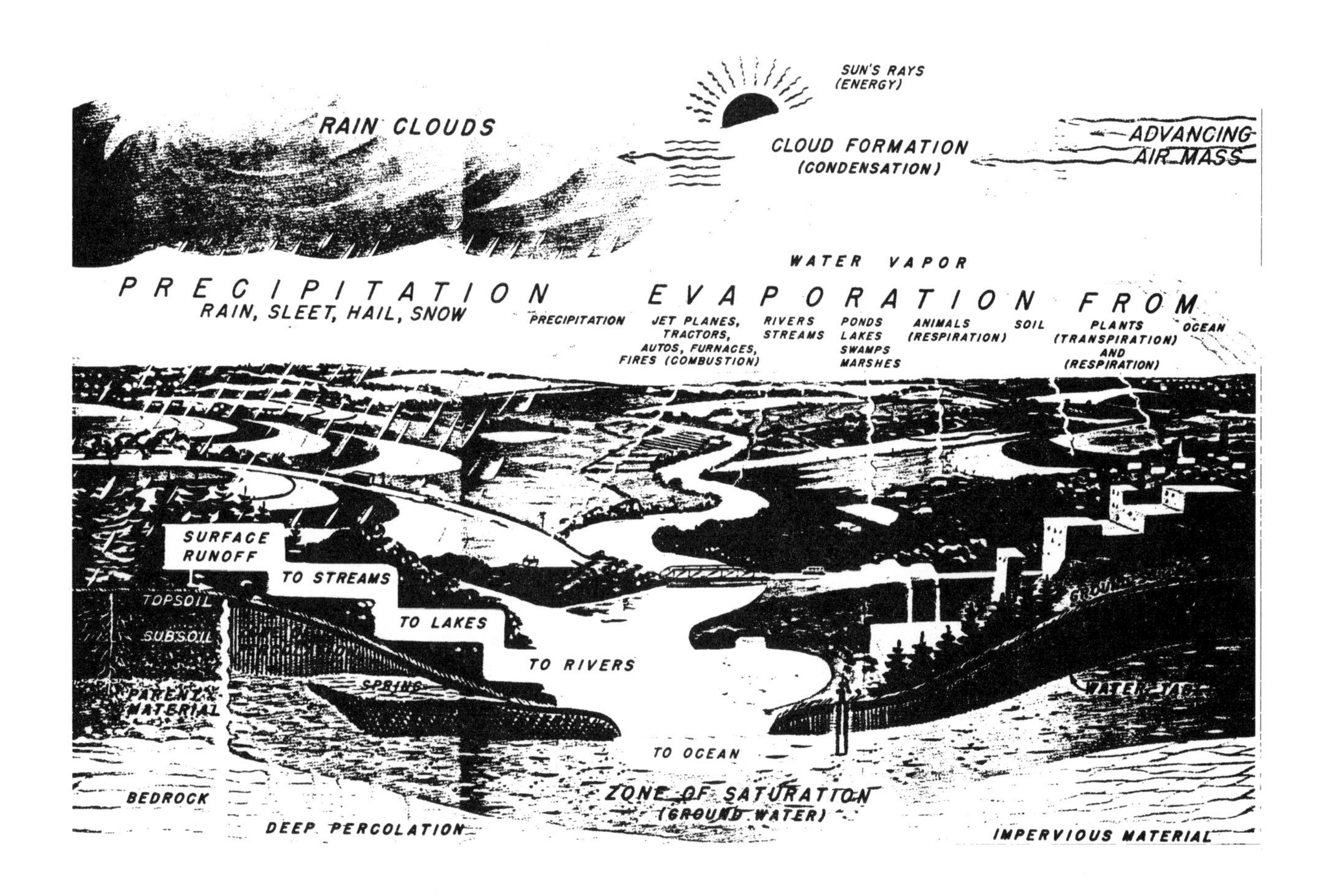

plants hold the soil in place and slow the runoff, allowing more water to soak into the soil.

Some of the water that soaks into the soil is used by plants. Part of it percolates beyond the reach of plant roots to the water table, to underground reservoirs, and to springs and artesian wells.

Runoff on its way to the sea can be intercepted and stored for industrial or household use, and it can be diverted for irrigation.

Little water has been added or lost through the ages. The water cycle prevails in all places and at all times with neither beginning nor end.

Conservation and the Water Cycle

United States Department of Agriculture
Soil Conservation Service
Agriculture Information Bulletin No. 326

Water is probably the natural resource we all know best. All of us have had firsthand experience with it in its many forms—rain, hail, snow, ice, steam, fog, dew.

Yet, in spite of our daily use of it, water is probably the natural resource we least understand. How does water get into the clouds, and what happens to it when it reaches the Earth? Why is there sometimes too much and other times too little? And, most important, is there enough for all the plants, and all the animals, and all the people?

Water covers nearly three-fourths of the Earth; most is sea water. But sea water contains minerals and other substances, including those that make it salty, that are harmful to most land plants and animals. Still it is from the vast salty reservoirs, the seas and oceans, that most of our precipitation comes—no longer salty or mineral laden. Water moves from clouds to land and back to the ocean in a never-ending cycle. This is the water cycle, or the hydrologic cycle.

Ocean water evaporates into the atmosphere, leaving impurities behind, and moves across the Earth as water vapor. Water in lakes, ponds, rivers, and streams also evaporates and joins the moisture in the atmosphere. Soil, plants, people, and animals, and even factories, automobiles, tractors, and planes, contribute moisture. A small part of this moisture, or water vapor, is visible to us as fog, mist, or clouds. Water vapor condenses and falls to Earth as rain, snow, sleet, or hail, depending on region, climate, season, and topography.

Every year about 80,000 cubic miles of water evaporates from oceans and about 15,000 cubic miles from land sources. Since the amounts of water evaporated and precipitated are almost the same, about 95,000 cubic miles of water are moving between Earth and sky at all times.

Storms at sea return to the oceans much of the water evaporated from the oceans, so land areas get only about 24,000 cubic miles of water as precipitation. Precipitation on the land averages 26 inches a year, but it is not evenly distributed. Some places get less than 1 inch and others more than 400 inches.

The United States gets about 30 inches a year, or about 4300 billion gallons a day. Total streamflow from surface and underground sources is about 8.5 inches a year, or about 1200 billion gallons a day. This is the amount available for human use—homes, industry, irrigation, recreation.

The difference between precipitation and streamflow—21.5 inches a year, or 3100 billion gallons a day—is the amount returned to the atmosphere as vapor. It is roughly 70% of the total water supply. It includes the water used by plants.

Man can exist on a gallon or so of water a day for drinking, cooking, and washing, though he seldom does or has to. In medieval times he probably used no more than three to five gallons a day. In the 19th century, especially in Western nations, he was using about 95 gallons a day. At present in the United States, man uses about 1500 gallons a day for his needs and comforts including recreation, cooling, food production, and industrial supply.

When water hits the ground some soaks into the soil, and the rest runs off over the surface. The water that soaks into the soil sustains plant and animal life in the soil. Some seeps to underground reservoirs. Almost all of this water eventually enters the cycle once more.

Man can alter the water cycle but little, so his primary supply of water is firmly fixed. But he can manage and conserve water as it becomes available—when it falls on the land. If he fails to do so he loses the values that water has when used wisely.

Water management begins with soil management. Because our water supply comes to us as precipitation falling on the land, the fate of each drop of rain, each snowflake, each hailstone depends largely on where it falls—on the kind of soil and its cover.

A rainstorm or a heavy shower on bare soil loosens soil particles, and runoff—the water that does not soak into the soil—carries these particles away. This action, soil erosion by water, repeated many times, ruins land for most uses. Erosion, furthermore, is the source of sediment that fills streams, pollutes water, kills aquatic life, and shortens the useful life of dams and reservoirs.

Falling rain erodes any raw-earth surface. Bare, plowed farmland, cleared areas going into housing developments, and highway fills and banks are especially vulnerable.

In cities and suburbs, where much of the land is paved or covered—streets, buildings, shopping centers, airport runways—rainwater runs off as much as ten times faster than on unpaved land. Since this water cannot soak into the soil, it flows rapidly down storm drains or through sewer systems, contributing to floods and often carrying debris and other pollutants to streams.

Grass, trees, bushes, shrubs, and even weeds help break the force of raindrops and hold the soil in place. Where cultivated crops are grown, plowing and planting on the contour, terraces, and grassed waterways to carry surplus water from the fields are some of the conservation measures that slow running water. Stubble mulching protects the soil when it has no growing cover. Small dams on upper tributaries in a watershed help control runoff and help solve problems of too much water one time and not enough another time.

Throughout the world the need for water continues to increase. Population growth brings demands for more water. Per capita use of water, especially in industrialized countries, is increasing rapidly.

It is man's management of the precipitation available to him that determines whether or not he has both the quantity and the quality of water to meet his needs.

It is man's obligation to return water to streams, lakes, and oceans as clean as possible and with the least waste.

Note that (probably because this is an old publication) the estimate of water use is considerably lower than those provided in other sources. A more recent estimate is that the combined industrial, agricultural, and personal use of water in the United States averages out to 10,000 L (2600 gallons) per person per day.

This reading is a reprint of the U. S. Department of Agriculture Information Bulletin number 326. Originally issued in 1967, it was reissued in 1985 (U.S. Government Printing Office: 1985 0-475-503: QL 3).

READING 14

Where Can People Get the Water They Need?

Water recognizes no national boundaries; it is a global resource. Obtaining enough fresh water to meet human needs is becoming a global concern. Less than 1% of the world's water is readily available for human beings to use. There is water in the atmosphere, on the Earth's surface, underground, and in **sea ice** and **glaciers**, but almost 97% of the world's water supply is the salt water of the oceans. Most of Earth's fresh water is frozen in the glaciers of Antarctica and Greenland.

In this century, the human population has become so large that we have begun to interfere with the water cycle. In some areas where people withdraw groundwater faster than it is being replenished, the local effects are obvious: Wells dry up or the land subsides, sometimes forming sinkholes large enough for houses to fall into. It is likely that human activity is causing some changes in the water cycle on a global scale as well. These changes may cause wide-ranging effects in coming years. Clearly, we must find much more effective ways of managing this precious resource.

◆Water vapor: Water in the atmosphere

Water vapor is a gas mixed with the air we breathe. While water vapor is invisible, we can sometimes sense the presence or absence of water vapor in the air. When the air contains large amounts of water vapor, our skin feels sticky, and we say the day is "muggy." On cold winter days, the heated air inside a house may contain so little water vapor that your nose and eyes feel dry and uncomfortable and your lips and hands crack.

Water vapor can be extracted from the air. When you sit outside on a hot summer day sipping an iced drink, the outside of your glass gets wet. The glass does not leak; water vapor in the air **condenses** on the glass, forming the droplets that you see. The amount of water vapor that the air can carry without loss by condensation depends on the air temperature. The higher the temperature, the more vapor the air can hold.

When water vapor in the atmosphere condenses into tiny droplets, clouds form. If the droplets become large enough, they fall to Earth as snow or rain. While precipitation is an extremely important water source for human beings, the total amount of water vapor at any one time in the atmosphere is very small. If all the water vapor in the atmosphere suddenly condensed and fell to Earth, the resulting rain would form a layer only about 2.5 cm (1 inch) deep.

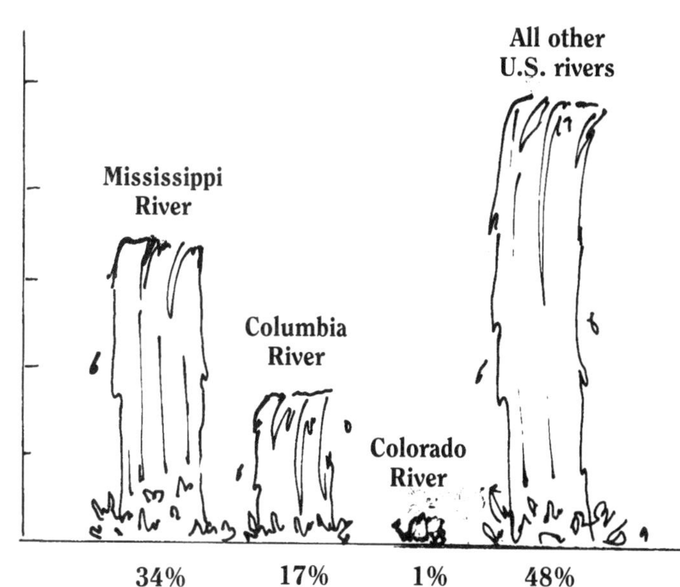

◆Surface water: The preferred water source

Lakes and rivers provide the United States with vast surface-water resources. In areas where surface water is available, does not contain large amounts of salt, and has not been polluted by human activity, it is the least expensive and most widely used water source.

The water in lakes and rivers is not evenly

distributed throughout the country, however. The Mississippi, North America's largest river, discharges 34% of all the river water flowing through the United States. The Columbia, our second-largest river, discharges about 17%; and the Colorado, a major water source for the dry western states, carries only 1% of all the river water flowing through the United States. All of these rivers are small in comparison to Brazil's Amazon, the largest river in the world, which has nearly 10 times more discharge than the Mississippi.

Lakes can be thought of as wide places in rivers. Earth's land areas are dotted with hundreds of thousands of small lakes. But these lakes hold only a minor amount of the world supply of fresh surface water. The combined volumes of Great Lakes and several other large lakes in North America comprise almost 26% of Earth's liquid, fresh, surface water. Even so, Lake Baikal in Asiatic Russia contains more water than the five North American Great Lakes combined. It is the largest and deepest single body of fresh water on Earth.

Not all lakes contain usable water, however. The water in many lakes contains too much salt or other dissolved minerals to be used for drinking or agriculture. Utah's Great Salt Lake is an example of this type of surface water resource. Worldwide, there is as much water in salty lakes as in freshwater lakes.

◆Groundwater: Vital to life

All the water beneath the surface of Earth is called, quite logically, **subsurface water.** Beneath most land areas of the world there is a zone where the pores of rocks and sediments are *completely saturated* with water. Hydrologists call water contained in the saturated zone **groundwater.** The volume of groundwater in the upper 800 m (one-half mile) of the continental crust is nearly 20 times greater than the combined volume of water in all rivers and lakes of the world.

The upper limit of the saturated zone of rock and sediment is called the **water table.** Groundwater lies beneath the water table. Water wells are simply holes that are dug or drilled deep enough in the ground to extend below the water table into the saturated zone. Water from the saturated

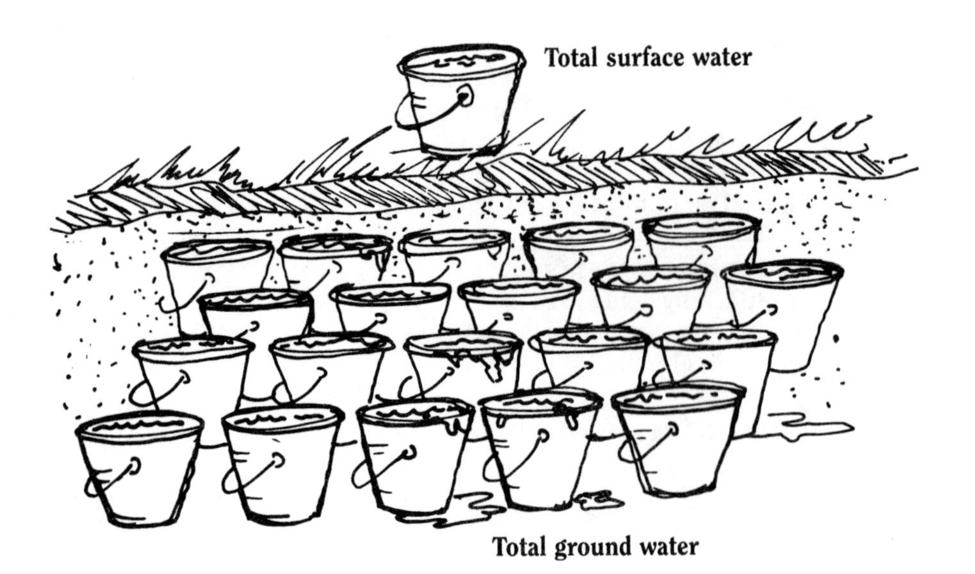

soil or rocks fills the well shaft; the water can then be drawn out for human use. The minimum depth of a well depends on the depth of the water table; wells are often very deep in arid areas, where the water table may lie hundreds of feet below the land surface.

Most of the plants that we use for food get the water that they need to survive from soil moisture contained in the zone of aeration of the soil *above* the water table. This soil moisture is classified as subsurface water rather than groundwater because it lies above the zone of saturation. Although the total amount of water held by soil above the water table is an insignificant percentage of the Earth's total water, this soil moisture is essential to all life. When the moisture in the soil near the surface becomes too low because of drought or insufficient irrigation, plants cannot survive; famine may result.

◆Ice: Our least-appreciated freshwater resource

Very few human beings make direct use of naturally-occurring ice as their primary water source; yet aside from the oceans, the greatest single item in the water budget of the world is the Antarctic. If the Antarctic ice were melted at a suitable uniform rate, its discharge rate would equal that of the Mississippi River for a period of more than 50,000 years. On the other hand, if a shift in climate led to rapid melting of polar ice, there would be a worldwide rise in sea level of over 50 m (180 feet), flooding all low-lying coastal areas and moving coastlines far inland.

Changes in the polar ice masses may also have indirect effects on the amount of available water. Scientists believe that changes occurring in the polar regions influence global weather patterns, therefore affecting the amount of rainfall reaching farmers throughout the world. Despite the scientific interest in polar ice masses, they seem remote and unimportant to the general population; we may drink from a stream whose source is a glacier, but few people consider how glaciers fit into the water cycle.

READING 15

Making Water Usable

◆Getting water to and from our homes

For many uses, such as drinking, citizens need water that is cleaner than that found in rivers and lakes. To clean or purify the water, cities and towns have built treatment systems, which cost money and require energy to run.

Water is pumped from a lake, river, or reservoir into the water filtration plant.

1. First, it is strained to keep fish and large objects out of the system.
2. Chemicals such as alum, chlorine (to kill bacteria), fluoride (to strengthen teeth), and lime (to prevent rust in water pipes) are added to the water at the *flash mixer.* Activated charcoal may also be added if taste and odor are problems.
3. The alum causes a chemical reaction in water that enables the dirt and other particles to stick together. This reaction is called **coagulation.**
4. Sticky, fluffy particles called **floc** are created by the alum, and during **flocculation,** dirt and other particles in the water are attracted to the floc and "clump" together.
5. In the **sedimentation** basins, the floc sinks to the bottom. This sediment, or solid matter, is called *sludge* and has to be removed from the plant. The disposal of sludge is a big problem for communities.

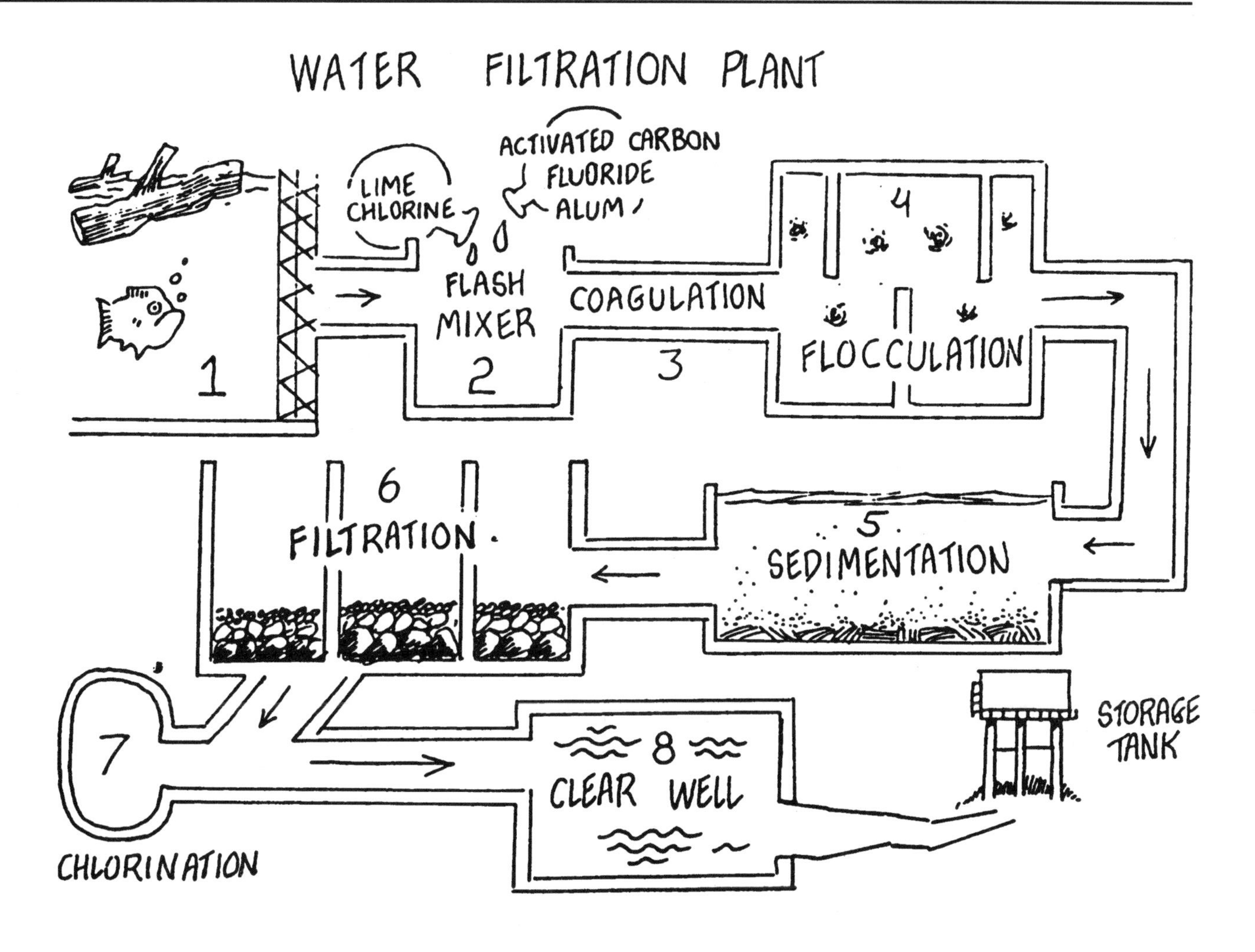

6. The clear water above the sediment is filtered through layers of sand and gravel to remove remaining dirt and other impurities (**filtration**).
7. Chlorine is added to kill any bacteria still remaining in the water (**chlorination**). Some plants may add fluoride and other chemicals at this stage.
8. The filtered, chlorinated water is stored in *clear wells* and in storage tanks until it is needed.

One way to remember the treatment process is to learn the "tion" words: *coagula*tion, *floccula*tion, *sedimenta*tion, *filtra*tion, and *chlorina*tion.

After the water has been treated, it's safe for our use. (Water can become polluted in the pipes or storage tanks before it reaches our homes, but this doesn't happen too often.) After we've finished using the water in our homes, it flows down drains and toilets as *wastewater.*

This reading is an excerpt from *Be Water-Wise*, produced by the Virginia Water Resources Research Center and reprinted with permission.

For more information about the series of *Be Water-Wise* instructional materials, contact:

Virginia Water Resources Research Center
Virginia Polytechnic Institute and State University
617 North Main Street
Blacksburg, Virginia 24060-3397
Phone (703) 961-5624

READING 16

Practical Tips for Conserving Water

◆Take me to your meter

First, turn off all water faucets and taps and don't flush the toilet. Next, find your water meter. It should look like one of these:

If any dial moves within a half-hour, you have a leak somewhere.

◆Drip...drip...drip

Check your water line connections and faucets for leaks. If necessary, tighten the connections.

Replace all the worn-out washers. You need only a screwdriver, pliers, and the right-size washer.

◆Beware of the strong, silent type

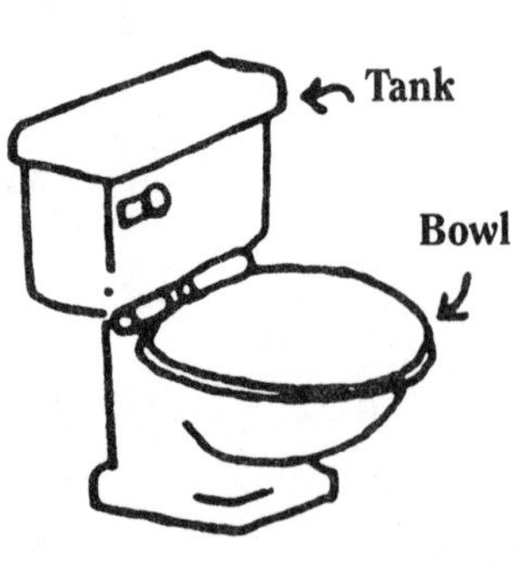

A toilet leak can waste hundreds of gallons of water a day. Listen for an ominous "HSSSS" sound. Since many leaks are silent, put a few drops of food coloring or a dye capsule in the toilet and wait 15 minutes. If the color shows up in the bowl and the toilet has not been flushed, you have a leak to repair.

Investigate new flushing devices that replace the ball cock and float. Many of these devices also have built-in leak detectors.

◆Sing shorter songs

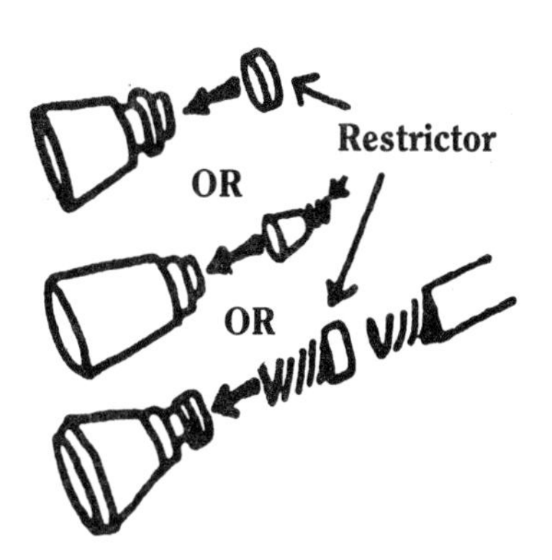

Shorten showers. A shower uses 5–10 gallons a minute. Use a kitchen timer as a reminder. Consider installing flow restrictors and water-saving shower heads. They are inexpensive and easy to install.

Turn off the water while shampooing or soaping up. A flow cut-off valve can be added to the shower head or purchased as part of the unit.

If you prefer tub baths, 1/4 of a tub should be enough. Put the stopper in the drain right away, rather than waiting for the water to warm up.

◆Toilet...water closet...commode

Whatever name you use, the toilet is the single largest water user in the home, accounting for 40% of a household's water use.

If it's agreeable to family members, flush only when necessary—two or three uses, or when there's solid waste.

Cut off the top of a plastic bottle, weight it down with some stones, and place in the toilet tank away from the flushing mechanism. When you flush, you save the amount of water equal to the volume of the stones.

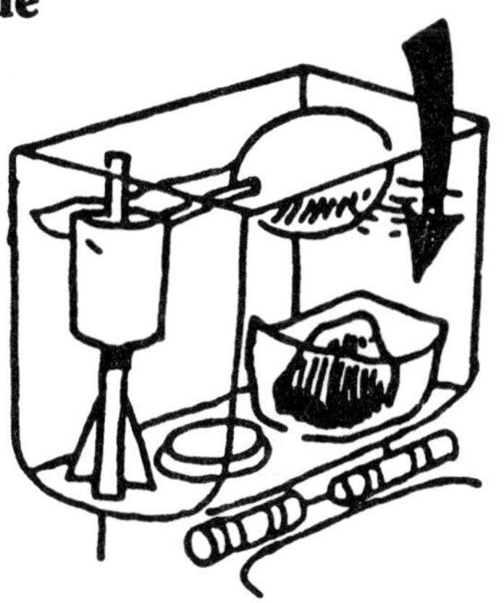

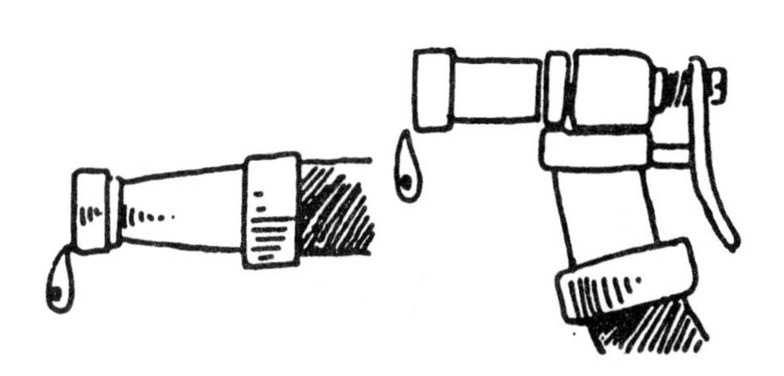

◆Close the hose

When washing the car, do not leave the hose running. Use a cutoff nozzle for easy shut-off.

◆To sprinkle or not to sprinkle

Lawns do not need regular watering. Less frequent waterings with sprinklers that spray low, broad drops will allow the water to seep into the ground, promoting deeper root systems that better withstand dry

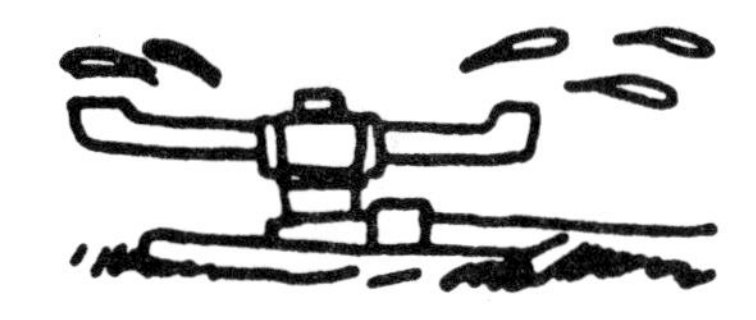

weather. Avoid watering too heavily because the soil cannot absorb too much water all at once and the extra will just run off. Leaving the hose running for four hours will use about 1500 gallons of water. Use a timer to avoid overwatering.

Mulch flower and vegetable gardens, shrubs, and trees to hold more moisture in the soil and to control weeds. In the garden, a soaker hose is the most efficient way to water because it puts water close to the roots and reduces evaporation.

◆Fill 'er up

In many washing machines, a full load of clothes uses the same amount of water as a half-load. Since most machines use 40–60 gallons of water, make every cycle count. Unless you load the machine to its rated capacity, you're not receiving full value from the water and energy you're using. The perma-press cycle uses 1/3 more water than regular settings.

If you've invested in a water-saving washing machine, remember to use the proper water-level setting.

Use dishpans or plug the sinks when washing dishes by hand. Don't let the water run continuously when washing or rinsing.

Load the dishwasher to capacity. Operating it partially-filled wastes water and energy.

◆Kitchen conservation

An inexpensive aerator attached to the kitchen faucet will save water.

Plug the drain or use a pan when washing vegetables. Later, the water can be poured on houseplants.

Use the garbage disposal sparingly. Accumulate the waste and dispose of it all at once by flushing with cold water—or better yet, save all the waste for composting.

Keep a jar of drinking water in the refrigerator, rather than letting the water run in the sink until

you get cold water. Or, to avoid opening the refrigerator door, keep ice water in a picnic jug on the kitchen counter.

Remember—

Your hands are the best conservation devices.

Use them:

To turn off the water when it's not being used, to fix leaks, to install water-saving devices, and more.

◆Be water-wise!

For more information, contact:

Virginia Water Resources Research Center
Virginia Polytechnic Institute and State University
617 North Main Street
Blacksburg, VA 24060-3397
Phone (703) 961-5624

This booklet, telling how to conserve water around the home, is part of the series *Be Water-Wise.* It is reprinted with permission from the Virginia Water Resources Research Center.

READING 17

What Is Water?

by H. Baldwin and L.B. Marman, Jr.

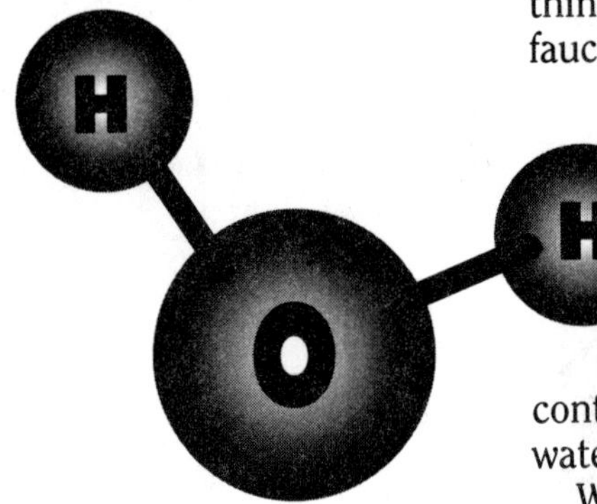

There is no simple answer to this question, for water means different things to different people. To most of us, water is what flows from a faucet, or what fills a pond or stream; it is the rain that makes the garden grow, or spoils a picnic. To a sportsman, water is a lake filled with fish, or a surface on which to sail, ski, or surf.

But no definition of water is complete without a discussion of its nature and its unusual properties.

Water makes up three-fourths of the Earth's surface and is the most common substance on Earth. Yet as common as it is, water is also the most precious substance on Earth. A contradiction? Not at all, for water is vital to life as we know it. In fact, water makes up two-thirds of our bodies.

Water's unusual properties, which make it so important to life, can be attributed to its remarkable chemical characteristics.

To a chemist, water is H_2O. This formula represents a molecule of water composed of two hydrogen atoms and one oxygen atom. Both molecules and the atoms that combine to form them are much too small to be seen with the unaided eye.

Water is actually more complex than its simple formula suggests. For our purposes, however, we need think of water only in less complicated terms.

The atoms in any molecule of any substance are joined together by a process known as chemical bonding. This bonding is particularly strong in water and results from the atoms' mutual need for more electrons that are parts of an atom.

To satisfy this need for electrons, water's oxygen atom "shares" each hydrogen atom's one electron with the hydrogen atom, and the hydrogen atoms each "share" an electron of the oxygen atom. The resulting chemical bond, coupled with other more complicated facets of water's chemistry, is the reason water has such unusual properties.

Strong chemical bonding accounts for water's remarkable ability to adhere to substances, that is, to wet them, and thus eventually to dissolve them. Chemical bonding also affects water's boiling point and its unique freezing process.

You are probably familiar with the three states of water found in nature: liquid, solid (ice), and gas (water vapor). Water is the only common substance on Earth that appears in all three of its natural states within the normal range of climatic conditions—sometimes at the same time. Familiar examples of water in its three natural states are rain, snow or hail, and steam.

As noted, water exhibits some unusual properties compared with other liquids. For example, most liquids contract steadily as they freeze. Water, however, contracts to a point but begins to expand as it reaches its freezing point of 0°C (32°F). This expansion can crack automobile engines or fracture rocks. It is an important part of the weathering, or breaking up, of rocks.

Because of this expansion, ice is less dense than liquid water. This is fortunate since, as a result, rivers and lakes freeze from the top down rather than from the bottom up. If freezing took place from the bottom up, some bodies of water might freeze solid, killing aquatic plant and animal life. Bottom-up freezing would also significantly affect our weather since many bodies of water in temperate parts of the country

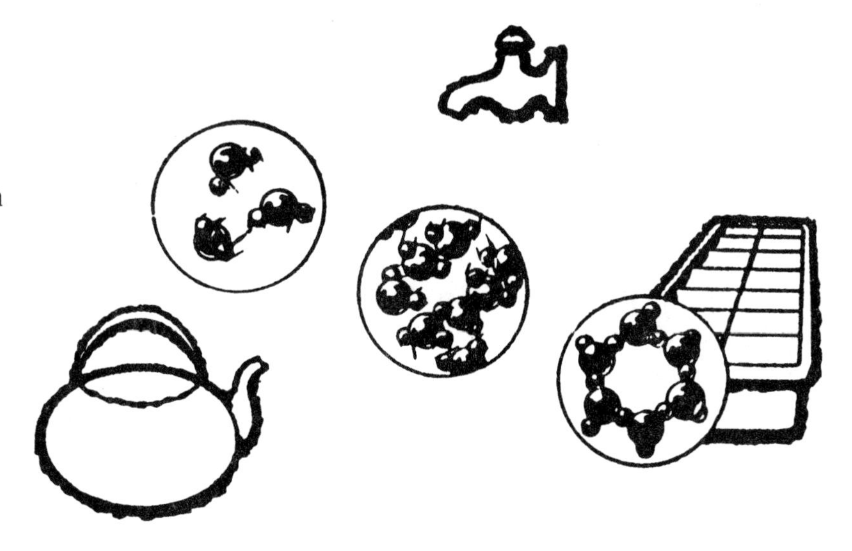

might never thaw completely, even in summer.

In ice, the water molecules are bound together in a nearly immobile crystal structure and the molecules do not move around each other. When ice is warmed to 0°C (32°F), it begins to melt and becomes liquid water.

As a liquid, the water molecules are less tightly bound together and can move around each other rather freely. The molecules' ability to slip and slide around gives water (and other liquids) its fluid properties.

As water becomes a vapor the situation becomes more complicated. Water boils at a temperature of 100°C (212°F) and becomes water vapor. Water can become vapor, however, at any temperature below its boiling point. Both ice and liquid water can evaporate into the air as vapor. Evaporation is part of the reason why puddles disappear after a rain. It is the water vapor in the air that gives you that "sticky" feeling on a hot, humid summer day.

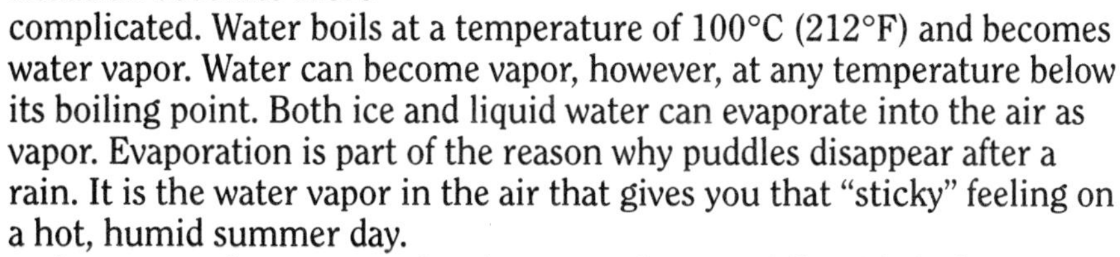

As a vapor, the water molecules move about rapidly with little attraction for each other.

Another unusual property of water is its *heat capacity,* that is, its ability to absorb heat without becoming extremely hot itself. In fact, water's heat capacity is second only to ammonia in nature.

Without water in it, a pan on a burner rapidly becomes red hot and then burns black. But put water in the pan and the water will absorb heat from it. The pan will become hot, but not as hot as before, and the temperature of the water, even if it boils, will rise only a small amount compared to the temperature of the pan without water in it.

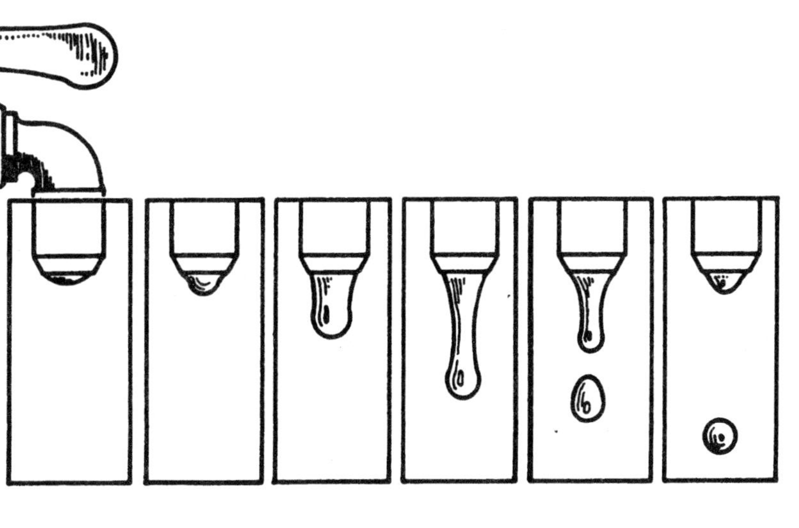

It is water's high heat capacity that helps make the oceans a key factor in the world's climate. The oceans absorb heat slowly during the day and during summer, and radiate, or give off, this heat slowly at night or during the winter. This process helps prevent our climate from extremes of heat or cold. Deserts, such as the Sahara in Africa, lack water's moderating influence and hence become scorching hot in the daytime and very cold at night.

Water also has a high *surface tension* in addition to its high heat capacity. Surface tension (or cohesion) is the ability of a substance to stick to itself and "pull itself together," or to cohere. As a drop of water falls from a tap it forms a sphere. A sphere is the shape that allows water to most closely pull itself together. Water's surface tension is so high that it is estimated that a force of 210,000 pounds would be needed to pull apart a column of water 1 inch in diameter. Since the water must be absolutely pure for this experiment, it is difficult to prove.

Its high surface tension enables water to support objects heavier than

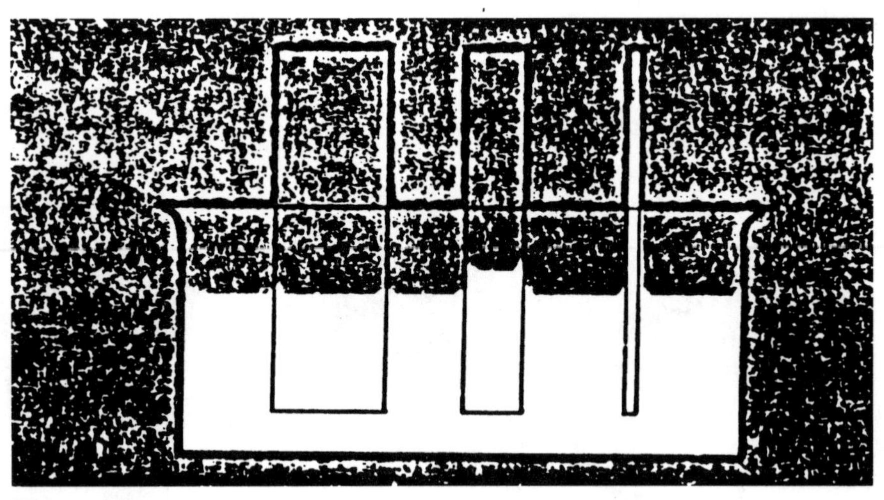
Water sticks

water itself, such as a sewing needle or insects that skate across the surface.

Water sticks, or adheres, to other surfaces as well. In a very narrow column such as a plant root or stem, the combination of surface tension and adhesion pulls water upward. This movement, known as capillarity, is partly responsible for the movement of water in soil as well as the movement of food in our bodies.

Perhaps water's most remarkable property is that given enough time, it can and will wear or dissolve away everything exposed to it. Examine a rock from a streambed and you will find it rounded and nearly smooth. This is partly due to water's action as an excellent solvent. The stream's flowing water has acted over the years to carry away small particles of the rock.

These processes whereby water gradually breaks down rocks into soil and then eventually carries even the soil away are called weathering and **erosion.** Water's ability to erode can often cause problems for man by carrying away fertile soil and later depositing (sedimentation) suspended soil particles (sediments) in reservoirs, ship channels, or other places where it is not desirable.

But where did all the water in the oceans, lakes, rivers, and under the ground come from?

Many scientists believe that as the Earth formed about four billion years ago its primitive atmosphere contained many chemicals that would have poisoned man had he been around. But among these chemicals were the basic gases needed to form water.

As the Earth cooled from a mass of molten rock, water formed in the atmosphere; then it began to rain. Scientists now believe that it rained for many, many years as the Earth continued to cool and the atmosphere we know today began to take shape. The rain came in such a large quantity that the low places on the Earth's surface were covered by water to a great depth, and the oceans were formed.

Ever since the rain began to fall and the oceans were formed, water has been trying to mold the Earth into a smooth surface. Other forces within the Earth keep raising up new hills and mountains; otherwise the Earth eventually would be covered by one vast ocean about 2400 m deep.

Scientific evidence indicates that life itself began in the primitive ocean and eventually found its way to dry land.

As the probable birthplace of life and because of its necessity to living things, life as we know it could not exist on Earth without that most abundant and remarkable of substances—water.

Adapted from the United States Department of the Interior/Geological Survey publication "What Is Water" by H. Baldwin and L. B. Marman, Jr. (U.S. Government Printing Office: 1981—341-618: 65).

READING 18

Why Is the Ocean Salty?

Generations of parents taking their children to the ocean have been perplexed by this question: What makes the ocean salty? After all, the rivers flowing into the ocean are fresh, and the rain falling onto the ocean is fresh, so why is ocean water 220 times as salty as the water in the Great Lakes?

by Herbert Swenson

All water, even rainwater, contains dissolved chemicals (dissolved minerals) that scientists call "salts." But not all water tastes salty. Water is fresh or salty according to individual judgment, and in making this decision man is more convinced by his sense of taste than by a laboratory test. It is one's taste buds that accept one water and reject another.

A simple experiment illustrates this. Fill three glasses with water from the kitchen faucet. Drink from one and it tastes fresh even though some dissolved salts are naturally present. Add a pinch of table salt to the second, and the water may taste fresh or slightly salty depending on a personal taste threshold and on the amount of salt held in a "pinch." But add a teaspoon of salt to the third and your taste buds vehemently protest that this water is too salty to drink; this glass of water has about the same salt content as a glass of sea water.

Obviously, the ocean, in contrast to the water we use daily, contains unacceptable amounts of dissolved chemicals; it is too salty for human consumption. How salty the ocean is, however, defies ordinary comprehension. Some scientists estimate that the oceans contain as much as 50 quadrillion tons (50 million billion tons) of dissolved solids.

If the salt in the sea could be removed and spread evenly over the Earth's land surface it would form a layer more than 500 feet thick—about the height of a 40-story office building. The saltiness of the ocean is more understandable when compared with the salt content of a freshwater lake. For example, when 1 cubic foot of sea water evaporates, it yields about 2.2 pounds of salt, but 1 cubic foot of fresh water from Lake Michigan contains only one-hundredth (0.01) of a pound of salt, or about one-sixth of an ounce. Thus, sea water is 220 times saltier than the fresh lake water.

What arouses the scientist's curiosity is not so much why the ocean is salty, but why it isn't fresh like the rivers and streams that empty into it. Further, what is the origin of the sea and of its "salts?" And how does one explain ocean water's remarkably uniform chemical composition? To these and related questions, scientists seek answers with full awareness that little about the oceans is understood.

◆Sources of the salts

Sea water has been defined as a weak solution of almost everything. Ocean water is indeed a complex solution of mineral salts and of decayed

If all the salt in the sea could be removed and spread over the Earth's surface, it would cover approximately one-half of the Empire State Building.

biologic matter that results from the teeming life in the seas. Most of the ocean's salts were derived from gradual processes such as the breaking up of the cooled igneous rocks of the Earth's crust by weathering and erosion, the wearing down of mountains, and the dissolving action of rains and streams that transported their mineral washings to the sea. Some of the ocean's salts have been dissolved from rocks and sediments below its floor. Other sources of salts include the solid and gaseous materials that escaped from the Earth's crust through volcanic vents or that originated in the atmosphere.

◆Why the sea is not fresh

The Mississippi, Amazon, and Yukon Rivers empty respectively into the Gulf of Mexico, the Atlantic Ocean, and the Pacific Ocean—all of which are salty. Why aren't the oceans as fresh as the river waters that empty into them? Because the saltiness of the ocean is the result of several natural influences and processes, the salt load of the streams entering the ocean is just one of these factors.

In the beginning the primeval seas must have been only slightly salty. But ever since the first rains descended upon the young Earth hundreds of millions of years ago and ran over the land breaking up rocks and transporting their minerals to the seas, the ocean has become saltier. It is estimated that the rivers and streams flowing from the United States alone discharge 225 million tons of dissolved solids and 513 million tons of suspended sediment annually to the sea. Recent calculations show yields of dissolved solids from other land masses that range from about 6 tons per square mile for Australia to about 120 tons per square mile for Europe. Throughout the world, rivers carry an estimated 4 billion tons of dissolved salts to the ocean annually. About the same tonnage of salt from the ocean water probably is deposited as sediment on the ocean bottom, and thus, yearly gains may offset yearly losses. In other words, the oceans today probably have a balanced salt input and outgo.

Past accumulations of dissolved and suspended solids into the sea do not explain completely why the ocean is salty. Salts become concentrated in the sea because the Sun's heat distills or vaporizes almost pure water from the surface of the sea and leaves the salts behind. This process is part of the continual exchange of water between the Earth and the atmosphere that is called the hydrologic cycle. Water vapor rises from the ocean

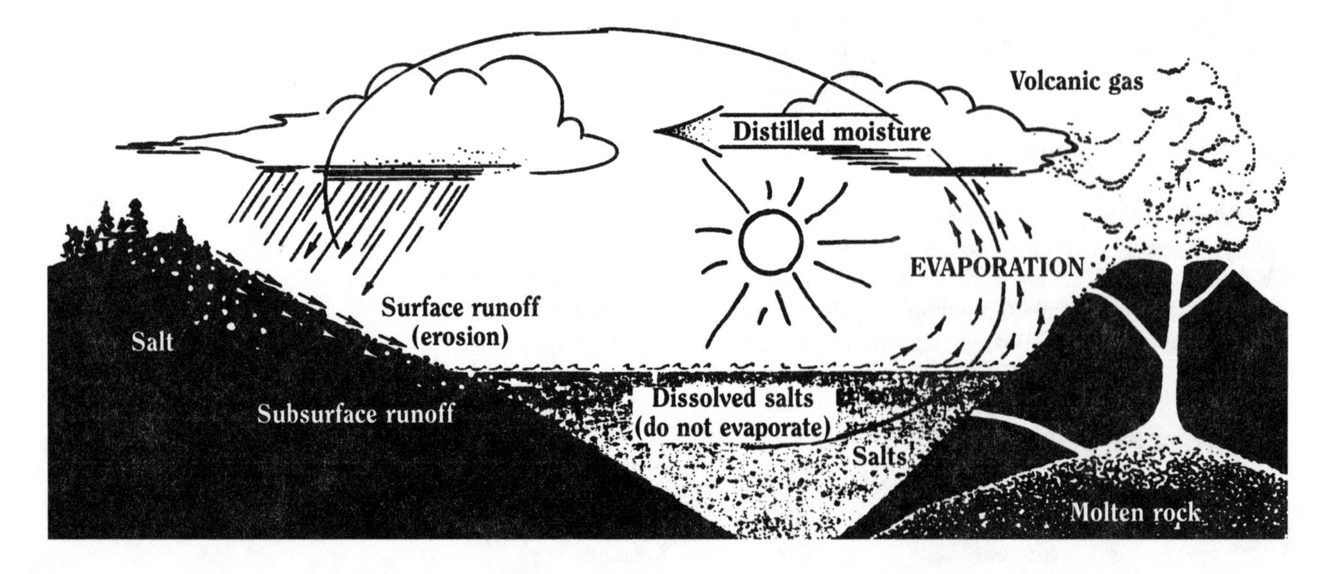

surface and is carried landward by the winds. When the vapor collides with a colder mass of air, it condenses (changes from a gas to a liquid) and falls to Earth as rain. The rain runs off into streams that in turn transport water to the ocean. Evaporation from both the land and the ocean again causes water to return to the atmosphere as vapor, and the cycle starts anew. The ocean, then, is not fresh like river water because of the huge accumulation of salts by evaporation and the contribution of new salts from the land. In fact, since the first rainfall the seas have become saltier.

◆Sea water is not simple

Scientists have studied the ocean's water for more than a century, but they still do not have a complete understanding of its chemical composition. This is partly due to the lack of precise methods and procedures for measuring the constituents in sea water. Some of the problems confronting scientists stem from the enormous size of the oceans, which cover about 70% of the Earth's surface, and the complex chemical system inherent in a marine environment in which constituents of sea water have intermingled over vast periods of time. At least 72 chemical elements have been identified in sea water, most in extremely small amounts. Probably all the Earth's naturally occurring elements exist in the sea. Elements may combine in various ways and form insoluble products (or precipitates) that sink to the ocean floor. But even these precipitates are subject to chemical alteration because of the overlying sea water that continues to exert its environmental influence.

◆How sea life affects sea water's composition

Inasmuch as the oceans receive most of their water from the rivers, the ratios (as distinguished from the total amounts) of different chemical constituents should be about the same in both regardless of total salt content. But this is not so. Comparisons of data on ocean water with the average salt concentrations in river waters of the world is as shown on the following chart.

Sea water and river water obviously are very different from each other: (1) Sodium and chloride (the components of common table salt) constitute a little more than 85% of the dissolved solids in ocean water and give to the water its characteristic salty taste, but they represent less than 16% of the salt content of river water; (2) rivers carry to the sea more calcium than chloride, but the oceans nevertheless contain about 46 times more chloride than calcium; (3) silica is a significant constituent of river water but not of sea water; and (4) calcium and bicarbonate account for nearly 50% of the dissolved solids in river water yet constitute less than 2% of the dissolved solids in ocean water. These variations seem contrary to what one would expect.

Part of the explanation is the role played by marine life—animals and plants—in ocean water's composition. Sea water is not simply a solution of salts and dissolved gases unaffected by living organisms in the sea. Mollusks (oysters, clams, and mussels, for example) extract calcium from the sea to build their shells and skeletons. Foraminifera (very small one-celled sea animals) and crustaceans (such as crabs, shrimp, lobsters, and

Comparison between ocean and river water

	Percentage of total salt content	
Chemical constituent	**Ocean water**	**River water**
Silica (SiO_2)	—	14.51
Iron (Fe)	—	0.74
Calcium (Ca)	1.19	16.62
Magnesium (Mg)	3.72	4.54
Sodium (Na)	30.53	6.98
Potassium (K)	1.11	2.55
Bicarbonate (HCO_3)	0.42	31.90
Sulfate (SO_4)	7.67	12.41
Chloride (Cl)	55.16	8.64
Nitrate (NO_3)	—	1.11
Bromide (Br)	0.20	—
Total	**100.00**	**100.00**

barnacles) likewise take out large amounts of calcium salts to build their bodies. Coral reefs, common in warm tropical seas, consist mostly of limestone (calcium carbonate) formed over millions of years from the skeletons of billions of small corals and other sea animals. Plankton (tiny floating animal and plant life) also exerts control on the composition of sea water. Diatoms, members of the plankton community, require silica to form their shells, and they draw heavily on the ocean's silica for this purpose.

Some marine organisms concentrate or secrete chemical elements that are present in such minute amounts in sea water as to be almost undetectable: Lobsters concentrate copper and cobalt; snails secrete lead; the sea cucumber extracts vanadium; and sponges and certain seaweeds remove iodine from the sea.

Sea life has a strong influence on the composition of sea water. However, some elements in sea water are not affected to any apparent extent by plant or animal life. For example, no known biological process removes the element sodium from the sea.

In addition to biological influences, the factors of solubility and physical-chemical reaction rates also help to explain the composition of sea water. The solubility of a constituent may limit its concentration in sea water. Excess calcium (more calcium than the water can hold) may be precipitated out of the water and deposited on the seafloor as calcium carbonate. Presumably as a result of physical-chemical reactions not well understood, the metal manganese occurs as nodules in many places on the ocean floor. Similarly, phosphorite (phosphate rock) is found in large amounts on the sea bottom off southern California and in lesser amounts in several other places.

◆Summary

The ocean is salty because of the gradual concentration of dissolved chemicals eroded from the Earth's crust and washed into the sea. Solid and gaseous ejections from volcanoes, suspended particles swept to the ocean from the land by onshore winds, and materials dissolved from sediments deposited on the ocean floor have also contributed. Salinity is increased by evaporation or by freezing of sea water, and it is decreased as a result of rainfall, runoff, or the melting of ice. The average salinity of sea water is 35%, but concentrations as high as 40% are observed in hot, dry areas such as the Red Sea and the Persian Gulf. Salinities are much less than average in coastal waters, in the polar seas, and near the mouths of large rivers.

Sea water not only is much saltier than river water but it also differs in the proportion of the various salts. Sodium and chloride constitute 85% of the dissolved solids in sea water and account for the characteristic salty taste. Certain constituents in sea water, such as calcium, magnesium, bicarbonate, and silica, are partly taken out of solution by biological organisms, chemical precipitation, or physical-chemical reactions. In open water the chemical composition of sea water is nearly constant. Because of the stable ratios of the principal constituents to total salt content, the determination of one major constituent can be used to calculate sea water salinity. For minor constituents and dissolved gases the composition is variable and therefore ratios cannot be used to calculate salt content.

Circulation and mixing, density and ocean currents, wind action, water temperature, solubility, and biochemical reactions are some of the factors that explain why the composition of water in the open sea is almost constant from place to place.

Adapted from the United States Department of the Interior/Geological Survey publication "Why Is the Ocean Salty?" by Herbert Swenson (U.S. Government Printing Office: 1984—421-618/116).

This publication is one of a series of general interest publications prepared by the U.S. Geological Survey to provide information about the Earth sciences, natural resources, and the environment. To obtain a catalog of additional titles in the series "Popular Publications of the U.S. Geological Survey," write Eastern Distribution Branch, U.S. Geological Survey, 604 South Pickett Street, Alexandria, VA 22304; or write the Western Distribution Branch, U.S. Geological Survey, Box 25286, Federal Center, Denver, CO 80225.

MATERIALS AND SOURCES

Guide for Teachers and Workshop Leaders

This section contains module-by-module master lists of materials, equipment, and audiovisual materials (if any) recommended for use in each workshop, and gives ordering information about the audiovisual materials recommended for use with this manual.

Master Lists of Materials

◆Module 1: Groundwater—The Largest Freshwater Resource in the USA

Master list of materials and equipment for each group:

1 dry, unglazed clay flowerpot
1 disposable metal pie pan
1 metric ruler
1 large paper grocery bag
1 large (1.5 L or larger) fruit- or vegetable-juice can
safety goggles for each group member
1 hammer
a 2 x 4 board about 35 cm in length
1 watch or clock with a second hand
1 waterproof marker, grease pencil, or dark crayon
7 small, disposable cups—clear plastic 250 ml (5 oz) cups are ideal
marbles or pebbles—enough to fill 2 of the disposable cups
fine, dry sand—enough to fill 2 of the disposable cups
1 eyedropper
2 dry soil samples—a "rich" and a "poor" sample
2 clear plastic disposable soft drink bottles (1 L size)
4 beakers—500 ml size (jars may be substituted)
scissors or knife to cut drink bottles
2 small (10 cm x 10 cm) squares of cloth
2 rubber bands
1 graduated cylinder (500 ml)
a water source

◆Module 2: Reshaping the Surface of the Earth

Master list of materials and equipment for each group:

1 stream trough* (each trough is made from a 101 cm [1.01 m] section of plastic rain gutter)
1 milk jug—3.8 L (1 gallon) size
2 pencils
1 collecting pan (a small rubber or plastic dishpan or bucket)
materials to serve as supports, such as bricks or wood blocks. You will need supports ranging from 14 cm high to 2 cm high (**Note:** These things may get wet).
1 meter stick
short sections of toothpicks (each about 1 cm long)
stopwatch capable of measuring 0.01 sec intervals
sediments: (1) sand (2) round pebbles (3) flat pebbles (4) powdered clay (china clay, kaolin, or pottery clay) (5) ion mixture (a saturated solution of table salt and water)
1 eyedropper
1 mixing jar (paper cup, baby food jar, etc.)
1 stirring rod (a pencil or plastic straw will work)
1 clean glass microscope slide
1 magnifying glass (or microscope, if available)
3 or 4 small (pea-sized) irregularly shaped pebbles
a spoonful of sand
2 plastic margarine tubs or plastic bowls of similar size
a common unglazed brick, brushed and rinsed to remove any loose particles
access to a freezer
a water source
rags, paper towels, or sponges for cleaning up

***Note:** Provide a bucket to collect all wet sediments. Do not dump sediments into a sink—they will clog the drain.*

Audiovisual equipment:

16 mm movie projector or
videocassette recorder and television

*Building stream troughs

The stream troughs used in Activities 7, 8, 9, and 10 may be assembled during the workshop by the workshop participants or ahead of time by the workshop leader. The following additional supplies are required to prepare the stream troughs and jugs used in these four activities:

sections of white plastic gutter for making stream troughs. The gutter is generally sold in 3.05 m (10 ft) lengths. You can make three 101 cm troughs from each section of gutter.
handsaw
waterproof marking pen
masking tape or duct tape
sandpaper or flat file
miter box (optional)
scissors with sharp point

◆Module 3: Raindrops Keep Falling on My Head

Master list of materials and equipment for each group:

safety goggles for each person
1 clear plastic soft drink bottle and its cap (use a 1 or 2 L bottle that is clear, not tinted green)
matches
1 light source—a flashlight or a small lamp
2 paint roller trays
2 plastic basins that are wider than the paint tray
1 sheet of notebook paper
1 pair scissors
1 watering can (a beach toy is fine)
topsoil—enough to partially fill the roller trays
mulch—grass clippings, straw, leaves, or shredded paper
several bricks or wooden blocks for supporting the roller trays
1 metric ruler or meter stick
newspapers or plastic sheeting
two 500 ml beakers or large paper cups
wide-range pH paper (either 0–14 or 1–13) and its color chart
one 100 ml graduated cylinder
two 250 ml (8 oz) cups
four 150 ml beakers (or 5 oz paper cups)
250 ml of white vinegar (5% acetic acid), commonly available in grocery stores
200 ml of clean silica sand (pure white quartz sand)
5 Tums™ tablets
waterproof marker
soda, lemon juice, vinegar
3.8 L (1 gallon) of distilled water (optional)

The following may be used by the entire class or workshop group:

collection bucket for the water containing sediments
a water source
rags, sponges, or paper towels to clean up spills

Audiovisual equipment:

16 mm movie projector or
videocassette recorder and television

◆Module 4: Water, Water Everywhere

Master list of materials and equipment for each group:

several empty 3.8 L (1 gallon) plastic milk jugs and/or buckets (as many as practical—each group needs at least 2 jugs, but having additional jugs and buckets is helpful)
one or more of each of the following sizes of graduated cylinders: 1000 ml, 100 ml, and 10 ml
other containers for dividing up "all the water in the world" (2 or 3 buckets or aquariums; 6 or 8 beakers or cups in a range of sizes—1000, 500, 250, and 100 ml beakers are recommended; 6 or 8 test tubes and test tube holders)
10 note cards or slips of paper to serve as labels
1 container for catching 1 L of water
a watch or clock with a second hand
a source of dripping water (use a sink, or you can make "leaky water sources" by poking a small hole in gallon milk jugs)

a container to catch the excess dripping water; for example, a bucket, cut-off bottom of a milk jug, or a pie pan
1 apple
1 strong plastic knife
several sheets of newspaper
1.5 L of "swampwater" (see "Preparation" section of Activity 21 for recipe)
3 empty 2 L plastic soft drink bottles with their caps
20 g of alum (potassium aluminum sulfate—approximately 2 tablespoons)
1 pair sharp-pointed scissors
fine sand (about 800 ml in volume)
coarse sand (about 800 ml in volume)
small pebbles (about 400 ml in volume)
a 5 cm square of flexible wire or nylon window screen
1 large (500 ml or larger) beaker (or a large jar)
tablespoon
rubber band

The following may be used by the entire class or workshop group:

two 122 L (32 gallon) trash cans
100 L of water
1 roll of masking tape
1 spring scale or laboratory balance
1 collection bucket for the water containing sediments
rags, sponges, or paper towels to clean up spills
an area in the schoolyard or parking lot that has a water source such as a garden hose (optional)

◆Module 5: Investigating the Physical Properties of Water

Master list of materials and equipment for each group:

an area of blackboard
1 electric fan
2 sponges
a container of water
an electric lamp with a 100 watt (or lower) bulb
4 paper towels
scissors
plastic food wrap (20 cm x 20 cm piece)
the top of a cardboard shirt (gift) box
1 flexible, transparent (or translucent) plastic cup
fine-grained dry sand—enough to fill the cup
a pie pan or dish with a rim at least 1 cm high
1 metric ruler
1 beaker—200 to 600 ml volume (or a measuring cup)
1 metal or plastic cup
1.5 m of cotton twine

The following may be used by the entire class or workshop group:

1 roll masking tape
a meter stick
stapler
a water source
rags, paper towels, or sponges to clean up spills

Audiovisual equipment:

videocassette recorder and television

Recommended Audiovisual Materials

Depending on the needs, interests, and abilities of your students, segments of the *Eureka!* video series and films from the AGI-EBE Basic Earth Science Program and NASA may provide useful supplements or alternatives to textbook or lecture presentations of the concepts presented in these modules.

◆*Eureka!* Video Series:

Eureka! is an animated video series produced by TVOntario (© 1981). It uses examples drawn from everyday experiences to demonstrate the behavior of matter in motion. The cartoons are useful for showing how the motions of molecules of water change as they are heated, and how these motions account for the changes of state (solid to liquid to gas) that water undergoes.

Many school districts already have copies of *Eureka!* available for use. Check with your district or state instructional TV personnel to see if yours is one of them.

For information on how to purchase copies of *Eureka!* contact:

TVOntario
Suite 206
143 West Franklin Street
Chapel Hill, North Carolina 27514
1-800-331-9566 or (919)967-8004 (in NC)

◆Encyclopaedia Britannica Basic Earth Science Program Series

The workshops also include two films from the Encyclopaedia Britannica Educational Corporation Basic Earth Science Program Series, produced in cooperation with the American Geological Institute:

"Erosion: Leveling the Land" (No. 2194)
"What Makes Clouds" (No. 4240)

While these films are some of the older ones in the series (produced in 1967), their excellent quality makes them well worth using. Expect a few chuckles about the clothing worn back then! Summaries of these two films are included in Activities 6 and 12.

Another film in this series that you may wish to use in your classes is **"Evidence for the Ice Age" (No. 4196).**

The following content summary is based on the study guide for this film:

Summary of Film Content: "Evidence for the Ice Age"

Many natural objects and landforms are anomalies that cannot be explained in terms of present-day geologic processes or conditions: huge haystack boulders (erratics) surrounded by small hills in areas that are otherwise flat; empty river channels that, even during the wettest season, are occupied by small creeks; slightly weathered, solid rock that is polished, grooved, or scratched as though something had been dragged over it; broad U-shaped mountain valleys not eroded by streams; and stray pieces of copper found far from the nearest copper ore deposit—in places where rivers could not have carried them. Glaciation can explain all of these phenomena.

A glacier is a massive sheet of moving ice that forms in places where snow accumulates faster than it can evaporate or melt away. Year after year, as snow continues to pile up, it begins to press on previously accumulated layers. Subjected to such pressures, the buried snow turns into ice and starts to move under its own weight.

If a glacier were to move through a typical V-shaped mountain valley cut by running water, it would carve the sides of the valley (as well as the bottom) and would create a U-shaped channel. Also, rubble carried by the glacier would scratch, polish, and groove the valley sides. There are many glaciated or U-shaped valleys in North America—evidence that ice once moved through many of its mountain ranges.

When a glacier melts, it drops the material it carries—boulders and other sizes of rock debris—onto the ground. Glacial meltwater creates lakes and drainage systems. When empty, these former lakes and drainage systems can be recognized by the huge lake plains consisting of horizontal sediments. By knowing the location of evidence for glaciation, we can determine the shape and extent of a former ice sheet.

For information on how to purchase or rent films in the AGI-EBE Basic Earth Science Program, contact:

Encyclopaedia Britannica
Educational Corporation
425 North Michigan Avenue
Chicago, Illinois 60611
1-800-621-3900

◆NASA films

To borrow NASA films (there is no rental charge; however borrowers must pay the cost of return postage and insurance) write to the appropriate address below.

• If you live in Connecticut, Delaware, District of Columbia, Maine, Maryland, Massachusetts, New Hampshire, New Jersey, New York, Pennsylvania, Rhode Island, or Vermont, write to

Bill Derrickson
Wallops Flight Facility
Visitors Center
Building J-17
Wallops Island, VA 23337

• If you live in Alaska, Arizona, California, Hawaii, Idaho, Montana, Nevada, Oregon, Utah, Washington, or Wyoming, write to

NASA Ames Research Center
Public Affairs Office, 204-12
Moffett Field, CA 94035

• If you live in Alabama, Arkansas, Iowa, Louisiana, Mississippi, Missouri, or Tennessee, write to

NASA George C. Marshall Space Flight Center
Public Affairs Office, CA-20
Marshall Space Flight Center, AL 35812

• If you live in Florida, Georgia, Puerto Rico, or the Virgin Islands, write to

NASA John F. Kennedy Space Center
Public Affairs Office
Code PA-EAB
Kennedy Space Center, FL 32899

• In Kentucky, North and South Carolina, Virginia, and West Virginia, write to

NASA Langley Research Center
Mail Stop 185
Technical Library
Hampton, VA 23665

• In Illinois, Indiana, Michigan, Minnesota, Ohio, and Wisconsin, write to

NASA Lewis Research Center
Office of Educational Services
21000 Brookpark Road
Cleveland, OH 44135

• In Colorado, Kansas, Nebraska, New Mexico, North Dakota, Oklahoma, South Dakota, and Texas, write to

NASA Lyndon B. Johnson Space Center
Public Information Branch
Film Distribution Library
Houston, TX 77058

When ordering a film, please give the name, address, and zip code of the person and organization assuming responsibility for the film.

Metric Conversions

Only two countries in the world, Burma and the United States, persist in using the English system of measurement.* Although some U.S. industries are beginning to build items to metric standard measurements in order to become more competitive in international markets, the old units of feet, pounds, quarts, and other English measures are still commonly used for many consumer items in the United States. Until we join England in giving up the English system, converting from English to metric units or vice versa will sometimes be necessary. The following table tells how to do the most common conversions:

Rules for English to Metric Conversions

Multiply	by	this number		to get
_____ inches	x	2.54	=	_____ centimeters
_____ feet	x	0.3048	=	_____ meters
_____ miles	x	1.609	=	_____ kilometers
_____ quarts	x	0.945	=	_____ liters
_____ gallons	x	3.7	=	_____ liters
_____ pounds	x	0.453	=	_____ kilograms
_____ °F – 32	x	.556	=	_____ °C

* *Popular Science*, Vol. 233, No. 1. (July 1988, p. 45, "What's News")

EARTH: THE WATER PLANET

Glossary

Acid: A substance that has a pH value between 0 and 7.

Acid deposition: Commonly called acid rain. It is water that falls to or condenses on the Earth's surface as rain, drizzle, snow, sleet, hail, dew, frost, or fog with a pH of less than 5.6.

Adhesion: The attraction between molecules that causes matter to cling or stick to other matter (as adhesive tape).

Aeration: The addition of air to water or to the pores in a soil.

Aquifer: An underground layer of rock, sediment, or soil that is filled or saturated with water.

Base: A substance that has a pH value between 7 and 14.

Capillarity: The process by which water rises through rock, sediment, or soil. The cohesion between water molecules and adhesion between water and other materials "pull" the water upward; also called capillary action.

Chlorination: The treatment of a substance, such as drinking water, with chlorine in order to kill disease-causing organisms.

Cloud: A mass of suspended water droplets and/or ice crystals in the atmosphere.

Cloud droplets: The tiny liquid pieces of water that many clouds are made of (some clouds are composed of tiny ice crystals). When cloud droplets join together and become heavy enough they form raindrops.

Coagulation: The process, such as in treatment of drinking water, by which dirt and other suspended particles become chemically "stuck together" (to form floc) so they can be removed from water.

Cohesion: The ability of a substance to stick to itself and "pull itself together."

Colloidal suspension: A method of sediment transport in which water turbulence (movement) supports the weight of the sediment particles, thereby keeping them from settling out or being deposited.

Condensation: Changing a gas into a liquid; for example, turning invisible water vapor into liquid water.

Condensation surfaces: Small particles of matter, such as dust and salt suspended in the atmosphere, which aid the condensation of water vapor in forming clouds; also called condensation nuclei.

Contour plowing: Plowing done in accordance with the natural outline or shape of the land by keeping the furrows or ditches at the same elevation as much as possible. Contour plowing reduces runoff and erosion.

Delta: Fan-shaped area at the mouth of a river (where seas are relatively calm). The deltas of large rivers such as the Mississippi extend many kilometers into the open sea.

Deposition: The process of dropping or getting rid of sediments by an erosional agent such as a river or glacier; also called sedimentation.

Dew: Moisture in the air that condenses on solid surfaces when the air is saturated with water vapor.

Dew point: The temperature at which the air becomes saturated with water vapor.

Discharge: The amount of water flowing past a location in a stream/river in a certain amount of time. Usually expressed in liters per second or gallons per minute.

Divide: A ridge or high area of land that separates one drainage basin from another.

Drainage basin: All of the area drained by a river system.

Erratic: Huge boulders, often carried for long distances by advancing ice sheets, that are dropped by melting glaciers and remain on the glaciated landscapes.

Erosion: The processes (including soil erosion) of picking up sediments, moving sediments, shaping sediments, and depositing sediments by various agents. Erosion plays a role in creating Earth's surface features—the landscape. Erosional agents include streams, glaciers, wind, and gravity.

Evaporation: Changing a liquid to a gas; for example, changing liquid water into water vapor.

Floc: Clumps of impurities removed from water during the purification process; formed when alum is added to impure water.

Floodplain: Area formed by fine sediments spreading out in the drainage basin on either side of the channel of a river as a result of the river's fluctuating water volume and velocity.

Fog: Clouds that form at the Earth's surface.

Frost: The ice that forms on surfaces as a result of the temperature of that surface reaching freezing before the air becomes saturated with water.

Geyser: A thermal spring that erupts intermittently and to different heights above the surface of the Earth. Eruptions occur when water deep in the spring is heated enough to turn into steam, which forces the liquid water above it out into the air.

Glacial striations: Lines carved into rock by overriding ice, showing the direction of glacial movement.

Glacier: A large mass of ice formed on land by the compacting and recrystallization of snow. Glaciers survive from year to year, and creep downslope or outward due to the stress of their own weight.

Groundwater: Water below the water table in the zone of saturation. Groundwater fills the spaces between soil and sediments or lies in the cracks and crevices in rocks. In a less preferred use, groundwater has the same meaning as subsurface water.

Gullying: Small-scale stream erosion.

Hail: Transparent or layered (ice and snow) balls or irregular lumps of solid water; hail usually falls from cumuliform clouds that have developed into cumulonimbus clouds or thunderheads.

Humus: Decomposed bits of plant and animal matter in the soil.

Hydrology: The scientific study of the behavior of water in the atmosphere, on the surface of the Earth, and underground.

Hypothesis: An informed guess.

Iceberg: Large chunks of ice that break off of coastal glaciers and float away.

Infiltration: The entrance or flow of water into the soil, sediment, or rocks of the Earth's surface; also called percolation.

Irrigation: Applying water by artificial means (not precipitation) to land used for agriculture.

Mulch: Material spread on the ground (Earth's surface) to reduce soil erosion and evaporation of water. Mulches include hay, plastic sheeting, and wood chips.

Mulching: Adding mulch to the ground (see above).

Percolation: See "infiltration" in this glossary.

Permeability: The capacity or ability of a porous rock, sediment, or soil to allow the movement of water through its pores.

pH: A relative scale of how acidic or basic (alkaline) a material is. This scale goes from 0 to 14; 7 is neutral. Acids have pH values less than 7; bases have pH values higher than 7.

Pore spaces: The open areas, or spaces, in soil, sediments, and rocks that are filled by air or water. If the pores are connected then water can flow through the material (the soil, sediment, or rock).

Porosity: A measure of the ratio of open space within a rock or soil to its total volume.

Precipitation: Water falling toward the Earth's surface in the form of rain, drizzle, hail, sleet, or snow.

Relative humidity: The ratio of the amount of moisture in the air to the maximum amount of moisture the air could hold under the same conditions; usually expressed as a percent.

Rills: Small grooves, furrows, or channels in soil made by water flowing down over its surface. Also another name for a stream—usually a small stream.

Runoff: Liquid water that travels over the surface of the Earth, moving downward due to the law of gravity. Runoff, which includes stream flow and overland flow, is one way in which water that falls as precipitation returns to the ocean.

Saltation: The movement of sand or fine sediment by short jumps above a streambed under the influence of a water current too weak to keep it permanently suspended in the moving water.

Saturation: The condition of being filled to capacity. When the atmosphere contains all the water vapor it can hold, it is filled to saturation, and the relative humidity is 100%.

Sea ice: Solid water that forms when ocean or sea water freezes. Sea water has an average freezing point of -1.9° C (28.6° F). However, the freezing point varies with the salinity of the water. Sea ice is much of the area around the Earth's North Pole.

Sediments: Fragments of material produced by weathering and erosion of rocks. Sediments may remain on the Earth's surface as soil, or may be transported and deposited in other locations by wind, streams, and other erosional agents.

Sheet wash: A flow of rainwater that covers the entire ground surface with a thin film and is not concentrated into streams.

Sleet: Precipitation that consists of clear pellets of ice; sleet is formed when raindrops fall through a layer of cold air and freeze.

Snow: Precipitation that consists of frozen flakes formed when water vapor accumulates on ice crystals, going directly to the ice phase.

Soil: Sediment on or near the Earth's surface that is formed by the chemical and physical weathering of rocks as well as the decay of living matter. Soil is sediment capable of supporting the growth of land plants.

Spring: Groundwater seeping or flowing out of the Earth's surface; springs occur where the water table reaches the surface.

Stream: The type of runoff where water flows in a channel downhill because of the pull of gravity. Streams include rivers, brooks, and creeks.

Sublimation: Formation of a gas from a solid, or vice-versa, without passing through the liquid phase.

Surface water: All water, fresh and salty, on the Earth's surface. Oceans, lakes, streams, snow, and glaciers are all surface water.

Suspension: A method of sediment transport in which air or water turbulence (movement) supports the weight of the sediment particles, thereby keeping them from settling out or being deposited.

Thermal spring: A warm- or hot-water spring; many occur in regions of recent volcanic activity and are fed by water heated by contact with hot rocks far below Earth's surface.

Till: A deposit of sediment formed at the base of a glacier, consisting of an unlayered mixture of clay, silt, sand, and gravel ranging widely in size and shape.

Topsoil: The top layer of soil. Topsoil can generally grow better crops partly because it has more organic matter (humus), allowing it to hold more water than lower soil layers.

Transpiration: The process by which living plants give off water vapor into the atmosphere.

Water cycle: The movement of water from the air to and below the Earth's surface and back into the air. The processes of evaporation, transpiration, condensation, infiltration, and runoff are all parts of the water cycle.

Watershed: A geographical portion of the Earth's surface from which water drains or runs off to a single place like a river; also called a drainage area.

Water table: The boundary in the ground between where the ground is saturated with water (zone of saturation) and where the ground is filled with water and air (zone of aeration).

Water vapor: The gaseous state of water.

Zone of aeration: The portion of the ground from the Earth's surface down to the water table. The zone of aeration is not saturated with water because its pores are filled partly by air and partly by water.

Zone of saturation: The portion of the ground below the water table where all the pores in rock, sediment, and soil are filled with water.